International Encyclopaedia of Agricultural Engineering

International Encyclopaedia of Agricultural Engineering

Volume 2

R. Maddala

ANMOL PUBLICATIONS PVT. LTD.
NEW DELHI - 110 002 (INDIA)

ANMOL PUBLICATIONS PVT. LTD.

H.O.: 4374/4B, Ansari Road, Darya Ganj,
New Delhi - 110 002 (India)

B.O.: No. 1015, Ist Main Road, BSK IIIrd Stage
IIIrd Phase, IIIrd Block,
Bangalore - 560 085 (India)
Visit us at: www.anmolpublication.com

International Encyclopaedia of Agricultural Engineering

First Published, 2007

ISBN 978-81-261-3375-8 (Set)

PRINTED IN INDIA

Printed at Mehra Offset Press Delhi.

Contents

Preface

This publication entitled "International Encyclopaedia of Agricultural Engineering" covers a wide spectrum of related subject matter, such as agriculture engineering, plant biotechnology crop production, plant agriculture, genetic engineering, plant genetic varieties, agro-ecosystems, aquaculture, commercial farming, crop-rotation, horticulture, sustainable agriculture, modern agricultural practices, seed technology, modern food industry, food-safety techniques, agricultural education, agricultural information system, crop-biotechnology, biodiversity, farming systems, global agricultural agenda etc. This wide-spread coverage of issues and themes makes this encyclopaedia a virtual reference book on the said subject. Besides, the encyclopaedia is designed in such a way that it creates understanding of the subject in a systematic manner, so as to greater a much required perspective about this much talked about subject.

The Encyclopaedia provides an introductory overview of agricultural engineering and biotechnology, with focus on the historical evolution of agricultural technologies. Both, agricultural biotechnology and microbiology, have been explained briefly, with focus particularly on fungal biotechnology. The states report on agricultural biotechnology at global level is presented, so are the details of various biotech tools being used to increase crop productivity. Their prospects and limitations in terms of agricultural production applications are covered in detail. Select case studies, such as cotton and maize farming, tobacco growing etc. are explained. Impart assessment of used of biotechnology in agricultural sectors, both as benefits and possible risks, discussed, including the issue of crop-yield sustainability. The relevant situation in India in this regard is also talked about briefly. This publication also presents a picture of the historical development of agricultural revolution and modern agricultural practices. This covers the Neolithic (first agriculture) revolution, green revolution and the role of high-yielding varieties (HYVs), and the Indian, British, Scottish agriculture revolutions. The main principles and practices of modern agricultural system globally are discussed, including processes like horticulture, corporate farming, crop-rotation, short rotation coppice (SRC), alternative crops in double-crop systems, disease resistance in fruits and vegetables, agricultural experiments station, best-agricultural practices and plan for modernization of present agricultural systems. The issue of seed technology, which has attained utmost importance today is also dealt in detail, so are the concepts and approaches of plant reproduction and growth. Special attention is given to various methods for promoting seed germination, with select case studies incorporated herein. Tools and techniques of contemporary agricultural technology and food industry and discussed, including their ecological impact. The focus lies on aquaculture technology, grain cleaners, grain and forage dryers, drying and aeration fans, chemical fertilizer applications, tractor performance monitors, potato planers, etc. An overview of agro-cultural education and extension services is given, with focus on agricultural engineering, technology and communication. Select case-studies are given to further elaborate upon newly developed machinery and automated operating systems in the field of agriculture, including the areas of pest and disease control. All this will help readers understand the academic linkages between various agro-systems, technology and education.

This Encyclopaedia also deals in detail the issue of crop biotechnology and modern agriculture. With particular reference to India. The much talked upon problem of food-safety at global level in also reflected upon briefly. The future of biotech feed crops for livestocks is also explained. The current status of biotechnology and its future challenges,, particularly in India, is dealt briefly. This publication covers extensively the subject areas of agro-ecology, sustainable farming, and role of agro-biotech, focusing mainly on the major benefits, risks and mythos. Select case studies have been mentioned to elaborate the point further. The issue of impact of Climate change and CO2 fertilization on crop yield is also dealt. Ethical questions being raised world-over against use of agricultural biotechnology and the myths surrounding the whole affairs have been addressed. Finally, this encyclopaedia also touches briefly upon the prevailing global agenda towards transforming the agricultural science policy. Relevant chapters of Agenda 21 and details of biodiversity action plan are presented. The Common Agricultural Policy and major policy recommendations for sustainable global agriculture are dealt in detail. An outlook is also given with regard to agricultural information management standards. Select case studies of workshops, conferences and publications are presented, so that readers can gain further insights into this areas of study, with focus on crop, forestry, fishery and animal husbandry sectors. An extensive bibliography, glossary is given, in addition to brief index.

—Editor

7

Crop Biotechnology & Indian Agriculture: Issues & Options

Development and History of Biotechnology

Science and technology is always an important component of human culture. Science directs technological innovation and technology accelerates the progress of science. Both are clearly distinct yet closely related.

Since the last century, the modern scientific and technological revolution has brought rich cultural achievements. Now, at this junction of the two centuries, we ask what will be the main trend of science and technology in the next century; what will their future look like; how should we evaluate the future of human culture? These are subjects both for futurology and for philosophy. In this article, I shall attempt a concise answer to these questions.

As the development of science is closely related to human culture, the different cultural backgrounds of Western and Eastern societies generated different paths for science. Western peoples through scientific exploration tried to understand the cosmic structure from the micro point of view, while Eastern people endeavored to discover the elementary structure of all matter from a macro point of view. From Shang to Han dynasties, China always led the way in scientific civilization. Why then did Chinese science fall behind after the seventeenth century when in the West a veritable Renaissance spread widely in Europe?

The famous physicist Li Zhendao in a speech at Fudan University in November 1992 said, "When we anticipate the 21st century, we must understand the principal problem in contemporary science. Only when these problems are fully understood is it possible that a breakthrough be made.... Once the principal problems are grasped, other problems can be readily solved." Thus a grasp of the principal problems in contemporary science can help to understand the development of science and technology in the century to come.

Whenever we try to recollect those outstanding scientists who made important contributions to the civilization of mankind since the Renaissance, and especially since the 19th century, the first to come to mind are Copernicus, Galileo, Newton, Faraday, Plank, Michelson and Einstein. In this list almost all are famous physicists. The helio-centric theory put forward by Copernicus and Galileo, the three laws of Newton's dynamics, Faraday's theory of electronic magnetism and Einstein's theory of relativity are all outstanding achievements in physics. Their application to technology resulted in the progress of human civilization. So we can conclude without hesitation that in the 19th and the 20th centuries the progress of civilization is owed to physics.

With the passing of time, more urgent and exact questions are presented to science and technology for still more advance in human civilization. Some problems that have puzzled humankind over a long period of time, such as why human beings must age and die, what is the nature of life and what is the essence of human being, have come up once again for discussion. This challenges the present technology, and the new problems cannot be solved by relying only on physical science and technology. The theory of evolution established by Charles Darwin in the 19th century has opened a totally new field for biology on the individual level and laid the foundation for the development of modern biology. But the new approaches to clarifying the phenomenon of life were not discovered until 1944 when the famous Austrian theoretical physicist and the founder of wave dynamics, E. Schrödinger, published "What is Life". This explained the essence of life through the theory of thermodynamics and quantum physics and elucidated the material structure of the organism, the sustenance of life and the heredity and variation of living things. This book played a major role in diverting the attention of some physicists to the problems of life science and urging biologists to investigate the mystery of life through the achievements of physics and chemistry. Some bold assumptions were made, especially the assumption that genetic material is a kind of organic macromolecule and that genetic characteristics were passed on through chromesomes in the form of a "code" —which later proved to be true. In the 1950's Click and Watson, under the influence of Shrödinger's theory and after careful analysis of the data concerning genetic material, suggested the double helix model for the structure for DNA. The establishment of the model marked a new age for the research of living materials on the molecular level.

The rise of molecular biology has brought new hope to the science of life. What exactly is life is the question natural science has long tried to answer. The development of molecular biology has led research on life in the direction of elucidating the basic structure and mechanism of the organism and has accelerated especially the advance of medical genetics. The achievement of molecular biology, along with computer science and modern physics and chemistry, can change our world greatly and can discover an effective way to solve many problems which concern humankind itself. Moreover, biotechnology, under the direction of molecular biology, will take on an entirely new aspect. For these reasons we can infer that biology will become one of the most attractive and promising branches of natural science in the next century.

The elementary purpose of life science is to reveal the origin of life, the mechanism of evolution, heredity and individual development, and the mystery of memory and thought. In essence, life science is deeply concerned with the development of humankind. A kind of driving force from both inside and outside is needed for the development of humankind and the progress of civilization. In the future this can be supplied by biotechnology. Hence, it can reasonably be imagined that the next century will focus largely on biotechnology as the future of civilization.

The Development of Biotechnology

The invention and application of biotechnology has a long history. In ancient times, whether in China or in foreign countries, people gradually selected various kinds of animals and plants to domesticate or turn into crops. Observing the variation of species, they began to notice different characteristics of the species and tried to keep these species at an excellent qualitative level. This can be regarded as the beginning of the application of biotechnology. In the middle of the 18th century the industrial revolution in the Western countries accelerated agricultural production. The appearance of Darwin's theory of evolution paved the way for recognizing the law of growth and variation of living beings. As a result the biotechniques of breeding came to be governed by exact and effective scientific theory. The experimental method was adopted to discover the pattern of variation, and the monk, Gregory Mendel, found two genetic laws by experimenting with peas. Mendel's work demonstrated that genetics had undergone the transition from phenomenal description to theoretical analysis and from empirical to theoretical science. Modern genetics and biotechnology owe much to Mendel's research in the last century.

Breeding in agricultural production is an example. Some method of breeding, such as sexual hybridization, seed selection and the creation of distinct varieties by physical means and chemicals were widely accepted soon after human society, for its livelihood, became engaged mainly in agriculture rather than hunting. These methods, in essence, were a kind of genetic engineering on the individual level.

With the advent of the 20th century, owing to the rapid increase of industrial production and the explosion of world population, the area under cultivation diminished. Moreover the aggravation of environmental pollution and the deterioration of the global climate has brought many unfavorable conditions to agricultural production. In some countries environmental pollution has caused soil erosion to thousands of hectares of good farmland. In Africa, the harsh climate constantly turns farmland to desert, making arable land more and more scarce. Obviously under such circumstances, the regular method of breeding cannot satisfy the basic requirement of human existence.

Though traditional biotechnologies do not lose the potential for developing new varieties, they have certain inevitable weaknesses. In the second place two species that are distant in evolution cannot be matched, so that the good characteristics are hard to transplant to one variety. Thirdly, repeated sexual hybridization results in a complexity of characteristics for the new generation which in turn will make the next hybridization more difficult. As for agricultural production, with the progress of human civilization, if a way to solve this were not found a crisis endangering the future of human society surely would spring up. Thus, the challenge of a better biotechnology was laid out before scientists.

A new technique of breeding, somatic cell hybridization, has been developed since the 1960's. To explain this simply, a specially devised procedure is used to cross-breed two or more cells from different species, so that the different chromosomes and genes co-exist in one hybrid cell; this hybrid cell is then transplanted to a normal host plant as a medium for growth into a complete and normal individual. Somatic cell hybridization is superior to sexual hybridization in the following aspects: 1. Somatic cell hybridization overcomes the restriction of non-affinity in sexual hybridization. 2. Rich varieties of cells ensure the availability of the material for cross-breeding. 3. The fusion of cells enlarges the sphere of cross-breeding. 4. The time needed for a new variety to grow into a mature plant is shortened. This implies as well the above wide adaptability of cell hybridization; its application will lead to a new agricultural revolution. This is a brilliant example of the combination of biotechnology with productive practice. This new biotechnique makes it possible to increase the yield of crops and to produce more nutritious food. It also can strengthen the resistance of crops to unfavorable climate, diseases and destructive insects. In addition, the rich varieties of cultivated crops can be retained and the gene resources consisting in the wild species conserved.

Nov adays, the crops created by somatic cell hybridization are very popular in a certain sphere and have produced good results. But with the success of somatic cell hybridization for plants, a new question has surfaced: can somatic cell hybridization be extended to animal cells? Some scientists have already striven to do so. Though there are still many technical difficulties, we can expect a brilliant future for this research which surely will have far-reaching influence upon the future of humankind.

Since ancient times, people have hoped for longevity, but have not been able to resist the natural law of aging and dying. In ancient China, many Taoist priests believed in alchemy and tried in vain to find a way of extending their lives. Tracing the real cause of their failure, we cannot help deriding their ignorance of science. The pills prepared by those Taoist priests contained heavy metal which is deadly. Now civilization has advanced; at the end of the 20th century science and technology are surging forward at a tremendous speed, and biotechnology has risen to new heights. At this time of transition the "task of investigating the cause of aging and death, which is deeply related to the progress of civilization, is put before biologists throughout the world. To this end, biologists are engaged in research using various techniques and so far have made encouraging progress.

For instance, some biologists believe that aging has much to do with genes. It is probable that aging is caused by a change of gene action when the human being reaches the late period of mature human reproduction. The cell-controlled genes which control the living beings are turned on and then off. The gene products responsible for differentiation and growth when they reach a certain concentration pass the information to special switches in the gene concerning reproduction, turning the switch on. In this way, the individual acquires the capability of reproduction. In the later period of reproduction, the excess of gene product results in a feedback to the gene controlling differentiation and growth so that the "switch" is turned off. This is probably the cause of aging. Though the theory is an imperfect one which needs to be proved by experiments, it has made encouraging progress in the efforts to solve the problem. Aging and dying are accepted as common sense, yet people still have never given up hope for longevity. The history of mankind suggests that with the advance of civilization and further adaptation to environment, we shall acquire higher flexibility and a stronger capacity for survival, so that life expectancy gradually and proportionally will be increased.

A great amount of data indicates that the average life span can surpass 100 years; the ancient hope for longevity is not a fancy. Biologists are making unremitting efforts to turn that hope into a fact by finding an effective way of postponing senility.

One more advanced technique is the growth factor, a kind of hormone with many usages, such as healing wounds and helping the growth of children. When injected into the human body it speeds up the metabolism so that fresh cells are produced more quickly and dead cells are discharged more effectively. This has many advantages: efficacy, convenience and minor side-effects. This method can be expected to become important in the next century.

Molecular biology has made great contributions to the advance of medical genetics. Since ancient times genetic diseases have been a difficult challenge for medical science. Modern biotechnology may serve as a promising tool to solve this problem by disclosing the nature of genetic diseases and providing prerequisite conditions for the treatment of these diseases. Modern biotechnology is able to focus on the ultimate cause of genetic disease—the chromosome where the variation has taken place. New technology helps to devise better procedures such as gene recombination to cure such diseases. At last the diagnostic technique for genetic diseases has been aimed directly at the detection of DNA instead of at the chromosome and protein. The technique of gene diagnosis has great clinical significance and it can be used to detect genetic disease with excellent accuracy, at great speed and without harmful effects to the patient; thus it has provided clinical medicine with an effective diagnostic method. In theory, genetic disease can be cured by gene transplantation, that is, to eliminate the abnormal variant from the cells of patients with genetic diseases and then put in normal genes. Though this therapy is still in the experimental stage with animals as its subjects, it promises a bright future. Gene therapy is a dream no more, but is waiting to be cultivated in order to cure patients so as to strengthen the potential to resist diseases and better adjust to the environment,

In sum, biotechnology has emerged under the impact of productive practice, with experimental research in molecular biology and genetics as its prelude. It not only has provided industrial and agricultural production with useful tools, but also perfects itself in the course of its application. This is the general trend of biotechnology.

The Future of Human Civilization

If we suppose that biotechnology will be the mainstream of technological culture in the next century, how will it influence the development of science and society? Of course, it can be only speculation, but it seems inevitable that the rapid development of biotechnology will lead to a great leap forward for human civilization, for advantages belonging solely to biotechnology can have immense impact on the future. These advantages are: 1. The recycling of the biological supply can provide the raw material so that there is no lack of resources; and 2. Low investment, a short period for generation, high economic benefits and wide utility of the products. All this suggests that biotechnology will be the leading industry in the next century and biological products will come to occupy the market. A dominant reason for biotechnology being so attractive is that it meets the requirements of the advance of civilization. The rapid development of science and technology and the rich varieties of the means of sustenance has made it possible for humankind to develop. Under such circumstances the rise of biotechnology brings hope for a solution of problems, for it can be used not only to modernize agriculture and industry, but also to solve environmental, ecological and health problems.

Science and man should never be placed in opposed positions. The more advanced science becomes, the more freedom humans acquire, so that civilization is driven forward. It was on the foundations of classical biology that modern biology was built. Biotechnology is an epistemological summary of Darwin's theory of evolution and the consequence of research in traditional physics, chemistry and biology. It established the goal of medical science. It results in progress for civilization as well as hope for the human future.

Of course, as all is dialectical in nature there is both a positive and a negative side. Hence, we must employ subjective initiative in order to amplify the positive side and to overcome the negative side and make it useful for humankind. We do not suggest a scientism which leaves the development of society out of account, and we oppose a humanism which inhibits the development of science. Only by associating science with the essence of man can we achieve its splendid future. The dialectic of the relation between biotechnology and man has the two sides, positive and negative. Antibiotics as one bioproduct possesses the remarkable ability to kill bacteria, but its improper administration may produce harmful effects on human health. For this reason there will be a thorough and systematic investigation before their complete acceptance. Thus a further demand is placed on biotechnology, namely, to advance only in the direction of benefiting mankind.

It should be pointed out here that contemporary scientists and philosophers have different views on the achievements of molecular biology. Some believe that modern biology is in conflict with social morals and ethics, and that the conflict cannot be compromised. They would engage the doctrine of the mean of the Confucian School and the theory of non-doing of the Taoist School. Close connection between Western and Eastern cultures on the basis of a correct understanding of the spirit of traditional Chinese culture can be beneficial. In contrast, to focus on the negative side and regard the problem as unsolvable will block the development of science and cannot be accepted. History is going forward, as does the human cognitive ability. Only by understanding the dialectical character of science and distinguishing what takes place of necessity from what can be controlled can we devise a splendid blueprint for the contemporary development of science.

Crop Biotechnology and Food Safety

Scientists and regulators have more detailed knowledge about approved biotech food crops – herbicide tolerant soya or insect resistant maize, for example – than any of the other fruit, vegetables, or animal products we eat regularly without thinking. Not only this, but hundreds of millions of people in North America have eaten these foods for the past six years, with not a single incident of harm. To quote the European Commission: “The use of more precise technology and the greater regulatory scrutiny probably makes them even safer than conventional plants and foods.”

On the other hand, when scientists are asked whether something is safe, they will never give an unequivocal, black-and-white “yes” as an answer. Nothing can ever be proved to be 100% safe. And yet, in a society which increasingly demands reassurance, where “accidents” become “preventable incidents” and where someone is always held to blame, such apparent lack of clarity makes some people worried.

Some of those concerned about the use of modern biotechnology in food production point to this inability to be completely certain as evidence that it is therefore unsafe and to be avoided. In reality, nothing could be further from the truth. Foods and ingredi- ents derived from genetically modified crops are the most stud- ied things we have ever eaten.

In this paper, we will look at some of the questions about safety, and the evidence produced to answer them. We will see that every application is thoroughly studied on a case-by-case basis before a decision is made by the authorities. We hope you will agree that this class of foods – highly regulated and exhaustively studied – presents no cause for concern. This is the fourth in a series of papers about aspects of modern biotechnology produced by ABE. We value your feedback and criticism, and would be pleased to have suggestions for further topics to be included.

The point has been made before that people's basic atti- tudes to new technology in general, and modern biotech- nology in particular, are remarkably similar across Europe and America. And yet there are clear differences in the reactions of societies as a whole to GM foods: very little real concern about their safety in the USA, and more positive attitudes in Spain, the Netherlands and Scandinavia than the EU on average, for example.

The major reasons for the differences observed on the two sides of the Atlantic are quite easy to see; in particular:

1. The United States has not had the series of high- profile food safety incidents experienced in Europe, BSE being the biggest.
2. Europeans in general, and many of the Member States in particular, have little confidence in their governments' ability to protect them from problems with food. We do not have the experience of an established, trusted EU-wide body equivalent to the Food and Drug Administration, although the setting up of the European Food Safety Authority will hopefully be the first step to correct this.

With this background, and in societies where environ- mental activists still command attention, it is easy to put doubts into people's minds about the safety of new and complex technology applied to something as per- sonal as food. Little wonder then, that, in consumer surveys, a significant proportion of Europeans express concerns about the safety of biotech foods

Food Safety: How is it Assessed?

A number of concerns have been raised about the safety of foods derived from modern biotechnology. The main ones, which we will cover in more detail below, are:

- Potential to induce new allergies
- Unintended effects, such as toxicity
- Possible transfer of genetic material to animals and people
- Antibiotic resistance marker genes

Allergies and their Causes

Allergies are caused by a range of substances: pollen, house-dust and food being the most common. In all cases, the allergy is to a particular protein, to which the body reacts. This reaction is, for most people, just annoying: a runny nose, itchy skin etc. However, for a few sensitive individuals, contact with even a very small quantity of the allergen (the protein responsible) can cause a severe reaction, called anaphylaxis or "toxic shock". In extreme cases, death can result.

The great majority of cases are caused by common items in most people's diets, in particular dairy prod- ucts, shellfish, soya and nuts. Certain allergens have a greater effect than others on susceptible individu- als; tree nuts and peanuts are a particular case in point.

A similar condition to allergy is celiac disease, in which wheat and related cereals causes inflammation of the digestive system and other problems. This is caused by a major component of the gluten protein in certain genetically-predisposed individuals. In the case of both allergies and celiac disease, sufferers have to steer clear of the food component they are sensitive to if they are to avoid all symptoms.

There is a feeling amongst the public at large that allergies are on the rise: a trend associated in many people's minds with modern living. For example, according to the British Nutrition Foundation, 20% of people believe they are allergic to one or more foods. In fact, based on a number of reports from round the world, the real incidence of food allergy is only 1-2 % in children, falling to less than 1% in adults. Of these, fortunately only a small percentage suffer from severe or even life-threatening reactions. In the case of celiac disease, the proportion of sufferers across Europe is estimated to be only about 1 in 5000, although there are locally higher incidences (up to 1 in 300 in parts of Ireland, for example).

Food intolerance (which is completely unrelated to aller- gy) is something which affects a larger number of people, although still well below 10% of the European population. It is a reaction – indigestion, for example – to a particular food which leads some people to avoid the food in ques- tion. However, this does not cause specific, severe reactions as allergies can, and is not life-threatening.

Could Biotech Foods Make a Difference?

Against this background, many people are understand- ably concerned about possible new allergens, and this is why some have questioned the safety of GM foods. Since genes code for the production of proteins, and proteins are potential allergens, doesn't the introduc- tion of different proteins create problems? This is exactly the question asked by the regulatory authori- ties when assessing the safety of a new food (whether or not produced by use of modern biotechnology).

Not surprisingly, there is a rational, stepwise approach to the problem

- The source of the inserted gene is, of course, known. If it is from a source known to cause allergies, the new protein must be presumed to be allergenic. This presumption can be confirmed by fur- ther testing, for example by doing skin tests on allergic individuals.
- Any new food shown to cause an allergic reaction in this way would not be allowed for human con- sumption.
- If the gene is not from a source commonly causing allergies, testing is more complex. For new pro- teins, the structure is compared with that of all known allergens to look for similarities. They are also subjected to a laboratory test for speed of digestion: those causing allergies are normally relatively slowly digested (since they must have time to cause a reaction). If the protein is dissimilar to known allergens, and is rapidly digested, it is unlikely to cause allergic reactions.

It is never possible to be 100% certain that something will not cause an allergy in some individuals: more than 160 foods have been associated with allergic reactions. However, we can be a lot more certain about newly- introduced biotech foods than other new additions to our diet. By way of comparison, many of us will remem- ber kiwi fruit first becoming available in Europe. No testing was required, and some individuals do indeed have a mild allergic reaction to them.

In practice, the regulatory regime has ensured that the public has not been exposed to new allergens from biotech food. The best-known example of this was a modified soyabean in development in the USA. This had a gene from Brazil nut added to increase the quality of the protein and improve its value as an animal feed. This was achieved, but at the laboratory testing stage it was shown that people allergic to Brazil nuts might also show a reaction to the modified soyabean. Because of this, development was stopped, and the improved soya never commercialised. It is now extremely unlikely that any company would choose to incorporate a gene from a known allergenic source in any new development. If – for whatever reason – such a gene was used and the benefits were thought likely to outweigh the risks, the proven regulatory processes would assure the safety of anything which reached the market.

Of course, in most cases, modern biotechnology will make no difference to existing allergies. Sensitive individuals will still have a reaction to some foods, whether modified or not. However, some groups of scientists are now work- ing on reducing the allergenicity of common foods using the tools of modern biotechnology. There is even the pos- sibility of a non-allergenic peanut being developed.

Further Safety Assessment: Looking for Unexpected Problems

Despite the scientific consensus on the basic safety of biotech foods, questions continue to be raised about potential long-term effects caused by regular consumption. Such potential problems can be assessed by feeding trials. However, it is one thing to test individual com- pounds (pharmaceuticals, for example) in this way, but how do we properly assess the safety of whole foods?

The safety of pharmaceuticals and other highly bio-active compounds is assessed by means of animal studies, using far higher quantities than would be eaten in practice, to establish a clear level having a harmful effect. Classically, a safety level is then set at 100-fold lower intake (on a body-weight basis) than the maximum level for which no effect was observed. However, in the case of whole foods, which are complex mixtures of tens of thousands of compounds, it is usu- ally impossible to feed high enough quantities to measure any effect.

In principle, a test diet containing artificially high levels of a novel protein could be devised, if this can be achieved without compromising the overall balance. Nevertheless, testing isolated components is not necessarily the same as eating the same component as part of a whole food, and care has to be taken to con- struct a meaningful experiment. Animals must be fed a palatable and nutritious diet, containing all necessary micronutrients. To assess the effect of a particular food component fairly, all other dietary components must be the same.

The difficulties associated with such experiments are covered in detail in the FAO/ WHO Expert Consultation report (2000). The Pusztai affair – covered in more detail below – is a perfect illustration of the way that a poorly-designed experiment can lead to incorrect conclusions.

In any case, there is a clear scientific consensus that it is simply unnecessary to do such tests in the large majority of cases. Where it is necessary, an example of how this problem can be approached in a meaningful way was reported by a Dutch group. They wanted to test the safety of tomatoes modified to produce a protein which controls certain insect pests (Bt tomatoes). First, they did a detailed analysis of all key nutrient and natural toxins and con- taminants. Then, they freeze-dried conventional and modified tomatoes and fed rats over a 90-day period with a standard diet containing 10% of the tomato powder. This was the equivalent of an average person eat- ing 13 kilos of fresh tomatoes every day! At the end of the test period, no adverse effects were found for either the trial or control group. If – in exceptional cases – the regulatory authorities deemed it necessary, it could be possible to take a similar approach for other foodstuffs.

For the major biotech crops now in commercial produc- tion (soya, maize and oilseed rape) there have been numerous feeding trials using more conventionally-formulated diets reported on both laboratory and farm animals. Since the primary use of these crops is to feed animals, the authorities required clear evidence of their safety before authorising their use, and further trials have also been done since. No adverse effects have ever been reported for approved biotech crops.

We should note again that testing requirements are determined by the regulatory authorities on a case-by- case basis, and that feeding trials are not necessary for all applications. We should also remember that hundreds of millions of ordinary people in North America and other countries have been eating these products on a regular basis for several years. There have been no cases at all of harm: surely the best evidence we have that these products are safe.

The Pusztai Affair

In 1998, Dr Arpad Pusztai, a respectedresearcher on plant lectins working at the RowettInstitute in Aberdeen, made statements on UKtelevision about his concerns over the safety of GM foods. These were based on experiments inwhich rats were fed raw potatoes, some of whichwere genetically modified. This was seized on byanti-GM campaigners as evidence that their con-cerns were scientifically justified. In the mean-time, Pusztai's results were reviewed, and he wasdismissed from the Institute. In the ensuingfurore, he became the hero of the activists, whoperceived him as a victimised whistle-blower onthe scientific establishment

However, when expert independent scientistsreviewed his evidence (made public on theInternet) they overwhelming came to the viewthat wrong conclusions had been drawn fromflawed experiments. For example, the animalswere fed a diet based on raw potato, which theycannot digest properly, and the modified potatohad a lower protein content than the control. Asimportantly, despite the tremendous publicity,no-one in succeeding years has been able to per-form any experiments in which animals sufferedany adverse effect when fed GM food tested andapproved under existing legislation.

The Royal Society, the UK's most senior andprestigious scientific organisation, set up a teamto evaluate Pusztai's evidence and claims, and inits report stated:

"We found no convincing evidence of adverse effects from GM potatoes. Where the data seemed to show slight differences between rats fed predominantly on GM and on non-GM pota- toes, the differences were uninterpretable because of the technical limitations of the exper- iments and the incorrect use of statistical tests."

The Lancet, the UK medical journal, eventually published a paper written by Dr Pusztai and his co-worker, Professor Stanley Ewen. This paper was defended by the editor of the Lancet as a way of putting the evidence into the mainstream of scientific debate. However, once again it was crit- icised by leading scientists. Sir Aaron Klug, then President of the Royal Society said:

"The Royal Society would not have published this paper by Professor Stanley Ewen and Dr Arpad Pusztai since it confirms the Society's original judgement that the experiments on which this paper is based were flawed."

Although this work will doubtless continue to be brought out as evidence of safety problems, the overwhelming scientific consensus is that there is no evidence to indicate any problems.

Can Genetic Material be Transferred to Animals and People?

Foods produced through modern biotechnology will not only have some (usually very minor) modification to the protein they contain, but will also have, by definition, some small change to their genetic makeup. We there- fore also have to consider possible safety aspects of this.

DNA (deoxyribonucleic acid) is the molecule which genes are made from. Whatever the gene or source, DNA is essentially the same: a long sugar chain, with four different bases attached in varying positions and proportions. The different patterns formed by these four bases are sometimes referred to as the "letters" of the genetic "alphabet". Combinations of these "let- ters" form individual genes which provide cells with instructions to make particular proteins. Because of the common structure of DNA in all life- forms, it becomes very difficult to say what is a "foreign" gene in a particular crop. A specific gene associated with a useful trait may first be identified in a certain organ- ism: an unrelated plant, for example. However, the genet- ic material which is incorporated in a modified crop plant is not used directly from that source. The particular DNA sequence is first copied thousands of times by incorpo- rating it in a harmless bacterium. The copies of the gene are then isolated and inserted into cells of the target crop to enable it to express the same trait.

Similarly, because all living things have ultimately evolved from the same early primitive forms of life, it is not surprising to learn that our genetic makeup has much in common with distantly related animals, plants and even bacteria. Basic cellular processes which define life are very similar, although different life forms have specialised in different directions: for example, unlike animals, plants photosynthesise but don't need muscles.

So, ultimately, DNA is DNA. Whether "natural" or mod- ified, the long chains are made up of exactly the same simple molecules. Animals (including humans) safely consume it every day. Our digestive systems treat it all the same: DNA is only biochemically active in cells. Once eaten, it is just another food component. In the words of the World Health Organisation: "The DNA from all living organisms is structurally similar. For this reason, the presence of transferred DNA in produce in itself poses no health impact to the consumers."

So, how much DNA is eaten, what happens to it in the digestive system, and what is the effect of modifying genes?

In the case of humans, a large proportion of our diet is cooked, and this process already denatures DNA. The only functional DNA we eat is in raw food: mainly fruits, vegetables and nuts. However, animals typically eat a diet of raw food, and therefore consume higher levels of native DNA.

By way of example, it has been estimated (see Beever and Kemp) that DNA represents about 0.02% of the dry matter of most crops, with every single cell containing about 10 10 -10 12 base pairs of DNA. A 600 kg dairy cow, fed on rations containing 60% GM maize, is estimated to 6 eat just over 600 mg of DNA a day (less than one gram). For a typical single gene insert (4000 base pairs), the total amount of "foreign" DNA in the diet would be only 0.0004% of the total: just 4 parts in 1 million.

Over 90% of DNA is fully digested by the range of enzymes in the guts of animals. Any fragments which cross the gut wall must be highly degraded, as large mol- ecules cannot pass this barrier. It is a normal, though minor, constituent of the diet of all animals, and there is no evidence that the genomes of humans or other ani- mals have ever incorporated functional DNA from food.

Don't forget that there is not only DNA in the food con- sumed by animals and Man. In the digestive systems of all animals, there is a vast range of bacteria, with their own particular genes. We are constantly exposed to this "foreign" DNA throughout our lifetimes, with no ill- effect or take-up of genetic material.

Antibiotic Resistance Marker Genes

When plant cells have new genetic material introduced into them (are "transformed") only a proportion of them actually receive and incorporate the introduced gene into their existing DNA. There has to be a way of finding out which ones have the new gene and which ones don't, before they are grown on as plants. In some cases, this is relatively easy: only cells modified to tolerate a par- ticular herbicide will grow in the presence of that com- pound, for instance. In other cases, the cells have to be distinguished using an additional "marker gene".

Genes commonly used in the development of the first generation of biotech crops included those which give resist- ance to particular antibiotics. Plant cells which had been subjected to treatment to introduce a new gene were then grown on media containing the antibiotic. The transformed cells, containing the antibiotic resistance marker gene (or AR marker gene) were able to grow, and non-transformed cells were not. So, the plantlets taken forward to grow on were known to have the desired additional gene.

From this point on, the AR marker gene serves no useful purpose, but remains part of the plant's genome. The question is then whether there are any possible negative effects from the presence of the gene or its products. In some instances, there have been concerns that the resistance might be transferred to gut bacte- ria which, in the presence of the antibiotic in question, could survive selectively. They in turn could theoreti- cally pass this resistance gene on to disease-causing bacteria, with possible health consequences.

In the rare cases that the AR gene is transferred to bacteria, this theoretical possibility only exists if the gene is still able to produce the protein conferring the particu- lar antibiotic resistance and is under the control of a bacterial promoter (ie, is able to function in bacteria). This is not the case for all applications of AR marker genes.

In practice, a large percentage of harmful bacteria in the environment are already resistant to one or more antibiotics. There are several reasons for this, not least over-liberal prescribing, patients failing to complete the course of treatment, and low-level use of antibi- otics in animal feed. This has led to the preferential selection of resistant bacteria, making some infections very difficult to treat effectively.

The theoretical risk associated with AR marker genes is very small indeed. They are, in the first case, chosen to confer resistance to antibiotics which are not commonly used in human or veterinary medicine. Then, the func- tional resistance gene must be transferred to bacteria in the gut of an animal (a rather unlikely event), and this gene must again be transferred in a fully working state to a bacterium harmful to humans. The potential additional hazard created by this is tiny compared to the present situation, but the EU authorities have taken the extremely precautionary position in the new legislation (Directive 2001/18) to phase out over the next few years the use of those AR marker genes which may represent a risk (how- ever small) to human health or the environment.

For an interesting insight into how one particular advi- sory committee approached this issue in a real case, see Professor Derek Burke's chapter in "GM Foods: understanding the issues".

Are there intrinsic problems with modern biotechnology?

Despite the enormous volume of work on safety assessment, there are still those who suggest that the very processes of modern biotechnology are in some way dangerous. One or two authors (see, for example, Mae-Wan Ho) have put forward theoretical concerns, in particular:

- So-called "promoter" sequences of DNA, incor- porated with genes to make them function properly, are often derived from plant viruses. It has been suggested that this can result in uncontrolled transfer of new genetic material into other plants, bacteria and animals. In fact, plants are very commonly infected with such viruses, which have evolved to enter cells in order to reproduce themselves. The additional transgenic sequences are identical to material which plants, animals and bacteria are exposed to in much higher levels on a continual basis. The "35S" promoter (derived from Cauliflower Mosaic Virus), the main object of the criticism, in any case does not function in bacteria.

- The process is also said to cause "genome instability", with sections of DNA rearrang- ing themselves in an uncontrolled fashion. Actually, the genome is not fixed and inflex- ible, but is subject to constant change in its natural state. The position of a particular gene in the genome can alter the way in which it works, but individual transforma- tion "events" are thoroughly characterised and checked for insert stability.
- The genome has a holistic nature, and incorpo- rating additional single genes will not have the claimed deterministic effect, but in some way disrupt the whole organism. This is clearly more of a philosophical than a scientific view.
- None of these criticisms has ever been consid- ered credible by biological scientists, and no experimental evidence has ever been put for- ward in support of the arguments. Of course, this doesn't mean we should be dismissive of legiti- mate concerns, but criticisms such as these do not have a scientific basis.
- The EU is a major sponsor of research, and this applies as much to biotechnology as other areas. In 2001, the European Commission published a report summarising the range of EU funded projects being undertaken o safety issues in the biotechnology area. This covers 81 projects, carried out over fifteen years (including those still in progress), involving over 400 research teams and costing some •70 million. As stated by Philippe Busquin, EU Research Commissioner, in his introduction:

  "The results of the research and growing practical experience, feeding into regulatory and risk man- agement policies, have enabled these to be regularly adapted to facilitate safe innovation, thus contributing to the excellent safety record to date, and providing a basis for continuing public confidence in the technology and its products."

The projects cover many safety-related issues, from detection methods to microbes used for cleaning con- taminated ground. In the specific case of food, nine major collaborative projects are listed. These cover such issues as improved standardised evaluation frameworks, possible gene transfer from foods to gut bacteria, and approaches to look at the unintended effects of gene insertion. Perhaps most importantly, there is a major project bringing together a wide range of safety assessment experts in a network designed to exchange information, discuss new issues of possible concern, and continually improve and share safety assessment protocols.

Reports such as this give the lie to criticisms that too little is known about safety issues. To quote from Professor Sir John Berringer's foreword:

> "One of the 'best kept secrets' during the last few years of acrimonious discussion about GMOs has been the enormous body of research being con- ducted in Europe and elsewhere that is directly rel- evant to risk assessments."

Of course, the fact that there is no evidence of any safety problems associated with food produced by agricultural biotechnology does not mean that we con- clude all applications are safe and that no further safe- ty assessment work need be done. For every new appli- cation under the Deliberate Release directive or Novel Foods regulation, the independent experts giving their opinion will raise specific issues and require evidence from well-designed, reproducible experiments to confirm that no problems exist. The particular data required is determined on a case-by-case basis. The continuing EU-funded research projects will improve still further the methods for collecting this evidence.

The Scientific Consensus

To date, there have been no serious doubts raised about the safety of biotech crops for ani- mals or people. Theoretical concerns have been expressed, but after careful study have never been given credence by expert scientists.

Numerous statements have been made by august scientific bodies in support of the technology: the EU Scientific Advisory Committees on Plants and Foods, the Royal Society, the US National Academy of Sciences and many others. Despite this, in today's society it is relatively easy to get concerns raised in the media, and the average per- son takes more heed of these than of the body of scientific information.

Public Perceptions: Now and in the Future

There are numerous reasons which contribute to the perceived public unease about biotech foods, some of which have already been covered above. The fact is that a significant proportion of the European popula- tion are suspicious of the use of modern biotechnology in their food 4. These suspicions often surface as con- cerns about safety.

And yet, the answer is not just to "educate" people about the facts. A full understanding of science and technology is to be encouraged, but we know that only a small proportion of the public will be suffi- ciently interested to get such knowledge. For the major- ity, it's a question of being reassured by those they trust. To put it simply, when the majority of people put more faith in government agencies than in environmentalists, the climate of opinion will be much more positive.

But government has to earn this trust, and building it is a slow process. The current approach at EU level is to tighten even further existing regulations in an attempt to address all the criticisms and concerns raised by the green lobby. Well thought-out, science-based legislation which inspires public confidence is clearly a good thing, but the present EU situation has the makings of a down- ward spiral. As new regulations are passed, further con- cerns are raised which are the subject of yet more pro- posed legislation. The reality is that those implacably opposed to crop biotechnology for political reasons will never be satisfied, and the result could be highly strin- gent regulations which stifle innovation without ever fully satisfying the critics.

Biotech Foods: The Most Thoroughly Studied Foods Ever

Foods derived from biotech crops have been studied more intensively and in more depth than anything else we consume. None have been released on the market without meeting all safety criteria devised by the relevant authorities. There have been no inci- dences of food safety problems attributable in any way to GM foods. This despite the fact that hundreds of millions of people, in North America in particular, have consumed these foods rou- tinely over the past six years as part of their normal diet.

We should also remember that enzymes (particularly chymosin for cheese-making and amylases for use in the starch industry) pro- duced from GM micro-organisms have been used for many years in an entirely safe and uncontroversial way.

In Europe (as well as America) the majority of farm animals regularly eat feed containing biotech soya. There have been no incidences of health problems for these animals. The products derived from them – meat, eggs, milk – are indis- tinguishable from those produced from non-GM fed animals.

To quote once more from the European Commission report: "The use of more precise technology and the greater regulatory scrutiny probably makes them even safer than conven- tional plants and foods."

Crop Biotechnology: Feeds for Livestock

Methods of plant breeding and biotechnology

Genetic modification of plants and animals has been practiced for approximately 10,000 years. By selecting the best plants and animals to serve as the parents for the next generation, our ancestors gradually modified wild plants and animals to create our familiar domesticated crops and livestock. Over the last century, new methods of plant breeding have been introduced, including wide crosses, F1 hybridization, mutagenesis, and induction of polyploidy. Biotechnology methods developed in the last few decades include tissue culture, embryo culture, anther culture, somaclonal variation, micropropagation, protoplast fusion, marker-assisted breeding, and transgenic plants (6). The terms "genetically modified" or "GM" are commonly used to describe these transgenic crops, but this is misleading because all domesticated crops have been genetically modified by one method or another. Thus, the term "biotech crops" will be used here to describe these crops developed using biotechnology methods.

Current & Future Biotech Feed Crops

Current Biotech Feed Crops. The biotech crop plants approved by USDA for commercial production in the US include canola, corn, cotton and soybeans enhanced with traits such as herbicide compatibility and protection against insects. Herbicide tolerance in biotech crops provides better weed control and more economical farm operations. Mangement of herbicide tolerant crops requires fewer agrichemical applications, so soil compaction and fossil fuel usage are reduced. Bt genes in biotech corn and cotton crops help to protect the plants against feeding by caterpillars of many moth species. A second benefit of the Bt genes is a reduction in the insect feeding wounds where fungal infection of the crop plant can occur. Many fungi that infect crop plants introduce mycotoxins into the feeds, producing a wide range of toxic effects in livestock including haemorrhaging, liver and kidney damage, diarrhoea, cancer and death. Monogastric animals such as horses and swine are generally more susceptible to mycotoxins than ruminants. By reducing fungal infection and mycotoxin content, Bt genes help to improve biotech feed quality and animal health. Enhancing the seed with pest protection traits reduces the need for chemical pesticides in biotech crop production, and often enables the grower to choose pesticides that are less harmful to the environment. Recent economic studies in the US and Canada document benefits of biotech crops including yield increases, cost savings to growers and reductions in chemical applications. These benefits have led to rapid adoption of biotech crops on over 100 million acres worldwide. Biotech crops will be planted on a large percentage of US acreage in 2001: corn (24%), cotton (64%), and soybeans (63%).

Feeding trials. Feeding trials with poultry, swine, fish, cattle and sheep have demonstrated that feeds derived from biotech crop plants are as safe and as nutritious as nonbiotech crops. Bt and nonBt corn hybrids had similar feeding value in beef steers, chickens and swine. Herbicide tolerant and conventional soybeans had equal feeding value in chickens, catfish and cattle. Biotech feeds are digested normally, so there is no effect on safety of meat, milk or eggs produced by the livestock, according to the Federation of Animal Science Societies.

Future Biotech Feed Crops. New biotechnology crop traits under development will provide additional benefits to livestock producers in the future. Potential future traits in biotech crops include more "output traits" that will enhance food and feed qualities. Biotech crop traits that may reach the marketplace within the next 5 years include corn with improved oil quality and improved protein quality. Alfalfa crop traits under development include enhanced protein quality and altered lignin for improved digestibility. Many of our traditional foods and feeds contain toxins and antinutrients such as cyanogenic glycosides, allergenic proteins, glucosinolates, phytate, glykoalkaloids, phenols, protease inhibitors and tannins. Biotechnology may help to reduce or eliminate these to produce more wholesome feeds. For example, biotechnology and plant breeding are being used to reduce phytate contents in corn and soybean seeds resulting in improved phosphorus uptake and utilization by livestock. Biotechnology could also be used to provide phytate digestive enzymes that might be added to the feeds to further reduce the phytate content and increase the feed value. The strategy of using digestive enzymes as feed additives has already been used successfully to solve a different feed problem in barley. The high content of glucan polysaccharides in barley and the inability of chickens to digest these glucans makes barley unsuitable as a feed for chickens. Biotech barley enhanced with a glucan digestive enzyme was fed to chickens and provided feed value equivalent to that of a corn diet. Vaccines delivered via biotech feeds may help to protect livestock against viral and bacterial diseases. This could reduce the need for antibiotics in feeds and for vaccine injections to protect the health of livestock. Biotech crops containing antifungal peptides may reduce fungal contamination in feeds. Together with the Bt gene strategy for protection against fungi discussed above, this will provide even better protection against mycotoxin contamination of livestock feeds.

Production of Biotech Feed Crops

Identity Preservation. Many customers are now setting stringent requirements in their purchase of biotech vs. conventional planting seeds and harvested products. Identity preservation is a system of crop management and detailed record keeping designed to document that distinct crop varieties have been kept separate from planting through harvest and on to the end user. As biotech crops with enhanced nutritional properties become available, identity preservation will be essential to ensure that the added value purchased is actually delivered in the feeds.

Crop Production — Pollen Drift. Pollen drift from biotech corn crops has been in the news recently, with concern about effects on monarch butterflies. Pollen from corn enhanced for insect protection has little or no effect on the monarch butterflies. Preliminary laboratory studies suggested that Bt genes and Cry protein expression in corn pollen might harm monarch butterflies. Recently more thorough field studies have demonstrated that biotech corn causes little or no harm to monarchs, especially compared to the insecticide spray treatment alternatives. One corn variety with high expression of Cry protein in pollen has been withdrawn from production. Corn varieties currently being grown have only minimal expression of Cry protein in their pollen.

Monarch populations have not declined in areas where biotech corn is being grown. The greatest risk to monarchs may be habitat loss, particularly in their overwintering sites.

StarLink Corn. StarLink corn contained a Cry9 protein (Bt gene) that was somewhat resistant to digestion in monogastric animals and therefore might act as an allergen in some people. StarLink corn is approved for animal feed and industrial applications. During planting and harvesting in 1999 and 2000 StarLink was not kept separate as intended, with some pollen drift into neighboring fields and some postharvest mixing in storage and transport. Thus, StarLink corn became mixed with corn intended for human food. Some reports of people with possible allergic reactions to StarLink are currently under investigation, but have not been confirmed. Registration for planting of StarLink has been withdrawn and planting seed for the 2001 corn crop is being carefully tested to ensure that no more StarLink is planted.

Government Regulation & Consumer Opinions

US Government Regulation of Biotech Crops. The use of crops produced via biotechnology is regulated by three US government agencies. The USDA Animal and Plant Health Inspection Service (APHIS) regulates field-testing of biotechnology-derived plant varieties. The US Environmental Protection Agency (EPA) is the lead agency for evaluating crop plants with pesticidal traits developed via biotechnology. The US Food and Drug Administration (FDA) Center for Veterinary Medicine is the primary agency regulating the safety of livestock feeds developed via biotechnology.

International Government Regulation of Biotech Crops. Export markets for biotech crops are complicated by wide variability in international government regulations and consumer reactions to biotechnology. Many governments are still establishing mechanisms to regulate biotech crops and food products. The European Union Parliament recently repealed their moratorium on approval of new biotech crops, but governments in individual countries also need to act before these markets will open up to new biotech crops.

Public Debate about Biotech Foods. There is now a worldwide debate about the use of biotech crops and livestock. Opponents of agricultural biotechnology raise issues such as: food safety, food labeling, environmental contamination, reduction in biodiversity and concern about possible unknown consequences. Some of the protest against biotechnology is motivated by broader issues in society such as concerns about corporate monopolization of food, globalization of the economy and reduction of imports to protect local farm economies. New government regulations have enabled alternative choices for consumers who wish to avoid food products produced from transgenic crop plants. USDA regulations for organic agriculture certification exclude the use of biotech crop plants or their feeding to organic livestock. An additional consumer choice is provided by recent FDA regulations for voluntary labeling of foods produced without biotech ingredients. Scientific, social, economic and ethical issues are all involved in determining consumer acceptance of agricultural biotechnology.

Consumer surveys. Two recent public surveys in the US demonstrated that a majority support continued research and expect biotechnology to provide benefits. When asked about food safety issues, very few people indicated any concerns about biotechnology in foods. Many people are uninformed about how biotechnology contributes to food production, so there is a challenge to provide information about this to the public. In the US consumers are generally optimistic about biotechnology and believe that government regulations are working to protect public safety. Careful stewardship of the technology at all steps in the production and marketing chain is critical in order to retain the trust of the public.

Crop Biotechnology: Present Status

Western Europe has become resistant to food produced from genetically modified (GM) crops. Almost all of the European supermarket chains have removed GM crops from their own brands. There have been extremely active protests from a number of non-governmental organizations, including Greenpeace, Friends of the Earth, and Christian Aid. The EU and its member states are backing down from the approval of GM crops. The protests are spreading around the globe to the U.S. and other actual or potential producer countries. The conference tried to identify some of the causes for this.

Sir Robert May (Office of Science and Technology, London; for detailed views of May and the Office of Science and Technology on GM crops, see www.dti.gov.uk/ost/ostbusiness/index.htm) pointed out a taxonomy of his concerns, the major ones being food safety, transfer of genes from crops to wild species through cross-hybridization, and the effect of crop husbandry on impoverishing the countryside. The general agreement from critics and proponents alike was that, although there is a need to be careful, there are no major concerns about food safety of GM products, and that the U.K., the U.S., and other countries have put into place effective regulatory procedures that are much more rigorous for GM than for non-GM foods. For some crops in some regions (but not for a whole range of others such as wheat, maize, and rice, over most of their global distribution), cross-hybridization of crops with wild species does occur. This occurs for both GM and non-GM versionsthe question is, does it matter? Clearly, research is needed to establish this, and some results were presented by Detlef Bartsch (Aachen University of Technology, Germany) on gene flow from sugar beet to wild beet populations (Bartsch et al., 1999; Pohl-Orf et al., 1999). His experience suggests that it is not easy to demonstrate that accession of transgenes by wild relatives would provide a selective advantage for the recipient in the wild.

Benedict Haerlin (Greenpeace, Berlin) outlined the reasons why Greenpeace opposes the introduction of GM crops. These include issues related to politics and to our scientific knowledge about the crops. He expressed concern that genetic engineering creates unprecedented new genetic life forms, that genetically modified organisms (GMOs) cannot be recalled, that DNA exchange is not understood, and that the long-term evolutionary effects of introducing GMOs were unknown. He proposed that we proceed on the basis of what is not known rather than what is, and invoked the strictest use of the precautionary principle. Although he did not wish to say that Greenpeace would never accept GMOs, the organization has such a long list of questions that he saw little chance of these being satisfied in a decade or two. How many of these questions can ever be answered without large-scale experimental field releases, which Greenpeace also opposes, is completely unclear.

He expressed the view that the problems of GM crops and agriculture should not be left for technocrats to solve. Rather, Greenpeace feels that the big picture is political. Its concerns are the industrialization of agriculture, the movement of people from rural to urban areas, unlimited growth rather than sustainable growth, public good versus private business, and genetic engineering allowing agrochemical companies to "mutate" into life science companies. However, it was not clear that GMOs have had much to do with many of these issues, most of which predate the technology. In contrast, I and many others think that GMOs have the promise to address some of the real concerns raised by Greenpeace and others.

Greenpeace clearly believes that the majority of the public is of its opinion. It was therefore welcome that Richard Braun (BIO-LINK, Worb, Switzerland) was present to describe the Swiss experience and the results of the one democratic vote on biotechnology, which resulted in a large majority favoring the continuation of biotechnology in Switzerland. This was due in large part to the scientists eventually waking up to the need to engage the public in a debate and even to take to the streets to demonstrate their convictions. Braun emphasized that the dichotomy between the apparent distrust of science and the continued improvement in human well-being needs to be countered by scientists making a more active contribution to the social debate.

Robert May pointed out that the practice of agriculture had often been bad news for the rest of the natural world, and that we need to take care that GM crops do not make this worse. Brian Johnston (English Nature, Tauton, UK developed this theme in more detail. He listed a range of concerns relating to GM crops that are cross-fertile with weed species or tolerant to wide spectrum herbicides. However, he emphasized that there is the potential to develop GM crops that would contribute to a more sustainable agriculture. The problem may turn out to be that developing such crops may not provide commercial profit and therefore would need to be developed with public money. Undoubtedly, there is concern that this powerful technology is being used to further commercial aims at the expense of the public good. As May said, the first Green Revolution was achieved by the non-profit sector, but this one is being achieved by industry. The challenge will be to see that the potential public benefits of the technology are delivered in both the developed and developing world, even where they do not provide an attractive target for commercial research. Although much discussed during the 3 d of the conference, this issue was not resolved. In any event, success is likely to depend on sufficient funds being made available to public sector research organizations and on technology transfer from multinational companies to the public sector.

No technology is perfect, and the benefits and drawbacks of every technology are relative. GM crops need to be judged against the conventional technology they replace. Ben Miflin (IACR-Rothamsted) summarized some of the first analyses of the effect of GM crops in cultivation in the U.S. Surveys over time and place do not make direct comparisons of GM and non-GM crops. However, growers of GM cotton containing a gene for an insecticidal toxin derived from *Bacillus thuringiensis* (Bt) use one-half as much insecticide as growers of conventional cotton. It has been estimated that total insecticide use in cotton in 1998 was around 1,000 tons less than that used before Bt cotton was introduced (James, 1998 Gianessi and Carpenter, 1999. In addition, Munkvold et al. (1999) and Windham et al. (1999) have shown that the grain levels of mycotoxins (aflatoxin and fumonisin) produced by fungi that infect the grain following damage to the ear by corn borers is far less in certain Bt maize lines. Miflin argued that, in ignoring these benefits and trying to ban such crops, the non-governmental organizations such as Greenpeace are selling the public short, and that their opposition to GM crops may have more to do with its publicity value in their broader political campaign.

Scientists are also not immune from neglecting to carry out the appropriate comparisons. The work of Losey et al. (1999) on the monarch butterfly was discussed a number of times. Both Ian Denholm (IACR-Rothamsted) and Ian Baldwin (Max Planck Institute for Chemical Ecology, Jena, Germany) pointed out that it was an all too obvious result: Bt toxins have been known to be toxic to Lepidopteran species for a very long time, and sprinkling these toxins (in maize pollen or some other form) over the monarch butterfly's food will cause a toxic response. The same would be true if the milkweed had been dusted with dead bacterial cells containing the Bt toxin (the insecticide used by organic farmers) or with conventional synthetic chemical insecticides. The questions are: (a) does it happen in the field and (b) which method of insect control is least harmless to wild life? Bt crops may come out best in a proper comparative study.

The introduction of GM crops has also led to entirely new approaches to the commercial launching of a pesticide product. The requirements of producers and farmers to develop and implement strategies to combat the development of insect resistance to Bt crops is, as Denholm argued, a very significant step forward in resistance management. Such a requirement is not in force for conventional insecticides nor for the use of organic Bt sprays (for recent developments in requirements for refugia, see. The development of new insecticidal principles and their deployment in crops offers further opportunities for overcoming resistance (Schuler et al., 1998).

Developing Countries

There was widespread discussion on the application of biotechnology in developing countries. Most people agreed with the recent conclusion of the Nuffield Council on Bioethics that "The moral imperative for making GM crops readily and economically available to developing countries who want them is compelling" Unfortunately, as one of the speakers pointed out, the invited speakers on this topic were middle-aged white men from the West. If another conference is convened, the organizers should address this lack of balance (for the views of a female African scientist, see Wambugu, 1999

It is not appropriate for the well-fed and rich West to push its capitalist, socialist, or environmentalist philosophies onto the developing world. As Luis Herrera-Estrella (Centro de Investigacion y Estudios, Guanajuato, Mexico), the one scientific speaker from a developing country, forcibly pointed out, "non-governmental organizations, representing environmentalist philosophies of the well-off, are prepared to sacrifice the needs of the people of the developing world in the name of `Nature,' pretty much in the same way as 400 years ago, the `conquistadores' sacrificed the lives of the Native Americans in the name of `God'." (Herrera-Estrella, 1999)

The facts were communicated well by several speakers. The world's population, which reached 6 billion in October 1999, is estimated to reach around 8 billion in 2020; the majority of this increase is likely to occur in the cities of the developing world. Currently, there is around 0.25 ha of agricultural land per person in the world; however, whereas a Kenyan uses 0.1 ha, a Canadian has around 1 ha. By 2020, the world average will be down to about 0.17 ha. Despite improvements over the last 50 years, some 800 million people have insufficient food, and over a billion people, particularly women and children, are affected by specific nutrient deficiencies (iron, vitamin A, etc.).

Although the world may be able to produce enough food, particularly if Eastern Europe and the former Soviet Union make full use of their agricultural resources, the problems of politics, poverty, and distribution are still likely to prevent the needy from being fed. As Willy De Greef (Novartis Seeds, Basel) emphasized, there needs to be a distribution of technical knowledge to enable and enhance production close to the centers of need. The Consultative Group on International Agricultural Research (CGIAR) institutes (http://www.cgiar.org/) have been very successful (but not sufficiently) in doing this in the past with the first Green Revolution. Gordon Conway (Rockefeller Foundation, New York) made his well-known case for a second Doubly Green Revolution (Conway, 1999 Conway and Toenniessen, 1999). This revolution would involve both more productive crops and the application of modern ecology in the form of integrated pest management to conserve natural resources and the environment. Biotechnology can play an important part in improving crop performance, and Conway provided a number of possible objectives, including improving photosynthesis to increase crop yield potential, improving stress tolerance (especially on marginal land), and improving nutrition (e.g. rice to provide vitamin A, see below). In this revolution, the input of the farmers themselves is crucial in the planning and implementation of the research.

There was general agreement that the needs of the developing world probably will not be addressed by commercial organizations. This is especially true for those crops that are of minor importance in the West. However, multinationals can play an important role in the transfer of technology to the CGIAR institutes or to national governments via organizations such as ISAAA (http://www.isaaa.org/). There are already success stories in which such transfers have taken place and some crops are already in commercial production, but more needs to be done (James, 1997. Conway also argued for companies not to enforce patent rights in developing countries and to rely, when possible, on plant variety rights. He congratulated Monsanto for shelving the development of the so-called "terminator" technology.

Scientific Developments

Crop Protection

Most of the GM crops in production at the moment have modified crop protection characteristics, chiefly protection against insects and from competition (herbicide tolerance). While there are some commercial virus resistant crops, there are further possibilities for improvement, especially in relation to the plant diseases that prevail in developing countries. Progress, particularly in using pathogen-derived resistance, was reviewed by Roger Beachy (Donald Danforth Plant Science Center, St. Louis; Beachy, 1997. A more difficult challenge has been to engineer resistance to a range of fungal pathogens. Nevertheless, considerable progress has been made recently in understanding the genetics of the interactions between fungi and plants. Jonathan Jones (John Innes Centre, Norwich, UK) described advances in the isolation of resistance genes and also their molecular structure (Hammond-Kosack and Jones, 1997; Noel et al., 1999; Parniske and Jones, 1999). Arabidopsis carries around 100 resistance genes, whereas crop plants may have two to three times this number. Many have been used in conventional breeding by introgression from wild sources. This is a slow process, and the durability of a single resistance gene is short when deployed in monoculture. The molecular description of the resistance genes should enable them to be moved more rapidly, either by marker-assisted breeding or by transformation into crops. It should also enable a range of different resistance genes to be assembled in different transgenic lines of the same cultivar so as to allow mosaics of resistance genes to be used within a single field. The understanding of the interaction of the Leu repeat regions of resistance genes with the recognition proteins of the pathogen, allied with techniques of gene shuffling and in vitro evolution, opens up opportunities to produce new resistance gene variants.

Whereas agrochemicals have been successfully used against biotic stresses, they have had little effect in protecting crops against abiotic stresses. There are indications that biotechnology may be more suited to achieve this goal. One of the widest ranging stresses in world agriculture is pH, with some 40% of arable land being too acidic and another 20% too alkaline for optimal crop production. Acidic soils lead to metals such as aluminum becoming toxic, and to nutrients such as phosphate becoming deficient. Herrera-Estrella described an approach to engineering stress resistance based on the observation that certain acid-tolerant plants excrete organic acids that chelate and trap aluminum in the rhizosphere (De la Fuente-Martinez and Herrera-Estrella, 1999; Herrera-Estrella, 1999). Generating maize plants that overexpress cytosolic citrate synthase leads to excretion of organic acids by their roots. In experiments using plants in pots, these plants were found to have enhanced tolerance to toxic concentrations of aluminum and an increased capacity for phosphate uptake. Although Herrera-Estrella has sufficient material to conduct field tests, he pointed out that such GM field experiments in Mexico have been blocked under pressure from Greenpeace and other environmental groups.

Water is probably the crop resource that is in shortest supply and this condition will worsen. In addition, the quality of the water used for irrigation will decline because of a greater salt load. Because plants need to have their stomata open to take up CO_2 for carbon fixation, they lose water continuously through transpiration. This water needs to be replaced by the uptake of water from the soil. Can plants be created that lose less water in times of water deficit and yet carry out photosynthesis and grow? Julian Schroeder (University of California San Diego, La Jolla, CA) discussed research elucidating the genetics of some of the steps in the opening and closing of stomata. Stomata are regulated by major signal cascades involving abscisic acid, cytosolic calcium, protein kinases and phosphatases, potassium channels, and farensyl transferase (Ichida et al., 1997 Allen et al., 1999). Several of the genes involved have been cloned, and transgenic plants have been made. The results hold out the hope that modifications in stomatal control may be made that would favor more efficient water use.

End-Use Qualities

The recent development that was most widely mentioned during the conference was that of the β-carotene-rich, yellow rice created by I. Potrykus, P. Beyer, and colleagues. Ingo Potrykus (Swiss Federal Institute of Technology-ETH, Zurich) described the science behind this advance and some others in his laboratory (Ye et al., 2000). Rice endosperm does not contain any provitamin A (β-carotene). Theoretically, four enzymes complete the pathway from geranylgeranyl pyrophosphate to provitamin A, and genes for these enzymes were isolated from *Narcissus* and *Erwinia*. These genes were combined in transgenic rice, and some stable lines with yellow endosperms were produced. Biochemical analysis confirmed that the color was due to provitamin A. This was present in sufficient amounts such that a typical Asian rice diet using these lines would provide the daily requirement.

Iron deficiency has also been approached by a multigene strategy in which genes for phytase, ferritin, and a Cys-rich metallothionin-like protein were transferred and expressed in rice endosperm. Current lines have around twice as much iron as the wild type. This modified rice now has to go through the normal biosafety and agronomic tests prior to release into Asian fields. These are significant hurdles. However, there may be more obstacles, as de Greef noted that the EU rice buyers, who buy only a small proportion of Thai rice, had notified Thailand that its rice risked rejection if any of it was found to contain GM material. Thus, the EU and the powerful multinational buyers of cereals are exerting pressure that could block the introduction of what he described as "one of the most significant biotechnological developments of the last decade." The major harvested products of plants are polysaccharides, particularly starch. Lothar Willmitzer (Max Planck Institute of Molecular Plant Physiology, Golm bei Potsdam, Germany) discussed approaches he and his colleagues have taken to modify carbohydrate metabolism. They have made transgenic plants that produce modified starches that might be useful in industry.

Producing such starches in plants removes the need for certain chemical modifications that have environmental side effects (Lloyd et al., 1999). However, to achieve this has not been easy and the detailed story shows many examples of transgenic plants that failed to behave in the manner predicted from the biochemical hypotheses on which the choice of transgenes were based (e.g. Veramendi et al., 1999). The research has also led to identification of proteins and genes in starch granule breakdown that were not known nor predicted by biochemistry. Biopharming is a term used to describe the use of transgenic plants to produce pharmaceuticals. Mitch Hein (EPIcyte Pharmacetical, San Diego, CA) described the production of antibodies in plants. The first success led to the synthesis in tobacco leaves of antibodies effective against dental caries. Subsequently, it has been possible to synthesize high-affinity monoclonal secretory antibodies that can prevent microbial infection in humans. The technology can thus be considered for other immunotherapeutic uses, especially in mucosal tissue (Ma et al., 1999). Plants could be particularly valuable as commercial production systems for antibodies that are needed in large amounts. Current pharmaceutical production is expensive and limited; the total Western hemisphere fermentation production capacity is 500 kg. If antibodies are to be used for prophylaxis, production would need to be in excess of 5,000 kg/year. Plant production systems could provide this, but the current limitation is devising suitable purification procedures. Hein identified a number of potential targets, such as contraception and sexually transmitted herpes.

Improved Transformation

Although there have been a range of commercially successful transgenic crops, the technologies used for transformation have been relatively crude. There is a need to improve the efficiency of transformation, to limit the presence of unnecessary genes in the products, to direct insertion to specific sites, and to give more control over when, where, and how much expression of the transgene occurs. Not all of these topics were addressed, but Nam-Hai Chua (Rockefeller University, New York) described improvements in transformation based upon the utilization of genes promoting endogenous hormone production under the control of chemical signals. One system uses the *ipt* (isopentenyltransferase) gene from the Ti plasmid of *Agrobacterium tumefaciens* to increase cytokinin levels, leading to the generation of shoots from transformed plant cells (Kunkel et al., 1999). However, these shoots retain the shooty phenotype and result in sterile plants. Using a construct in which the *ipt* gene is only active in the presence of a chemical inducer, the transformation and early cultivation of the cultures is carried out in the presence of the inducer. When shoots have formed, the inducer is removed and whole fertile plants can be recovered with high efficiency. The use of chemical induction cassettes in conjunction with the CRE-lox recombination system (Ow, 1996 has allowed Chua and colleagues to trigger the removal of transferred DNA from transgenic plants. This has considerable potential for crop biosafety through the ability to remove transgenes, which are no longer needed, before the product reaches the fields and the market.

Pal Maliga (Waksman Institute, Rutgers University, Piscataway, NJ) pointed out that using the plastid genome as a site for transgenes has several advantages over the nuclear genome. These include transfer of the genes only via the female line (for many but not all plant species), thus preventing the movement of the transgene to wild species via pollen; very high levels of expression (so far up to 20%-25% of total cellular protein); and targeted homologous recombination into the plastid genome. The technology is challenging and only works routinely with tobacco, although Maliga is optimistic that it will soon work well in rice (Maliga and Nixon, 1998; Khan and Maliga, 1999).

Markers and Crop Improvement by Non-Transgenic Methods

Recombinant DNA technology has given rise to a range of methods that allow the genome to be tagged with DNA markers (Karp et al., 1997; Davis et al., 1999; Vuylsteke et al., 1999). Some of these methods and the new developments that may be expected, were reviewed by Mark Zabeau (University of Gent, Gent, Belgium). The crucial drive is to provide systems that can be automated using robots, chips, sophisticated analyses (e.g. MALDI-TOF), and computers so that large populations of plants can be handled in a cheap and routine manner. Based on the discussions at the conference and elsewhere, it would seem that considerable success is on the way. Such technology has a number of uses, the most important being the identification of agronomic trait loci and their movement into adapted cultivars using marker assisted breeding. This method of crop improvement is not (currently) subject to criticism and even received the vocal approval of Haerlin. Once loci have been identified, map-based cloning can be used to isolate the genes involved. Marc van Montagu (University of Gent) pointed out that this could be of value to mine for genes of importance (e.g. those responsible for the synthesis of pharmacologically active compounds) from exotic species.

The practical value of marker technology was exemplified by the talk of Susan McCouch (Cornell University, Ithaca, NY), in which she described an experiment done with rice involving scientists in many centers in different countries, which was designed to discover quantitative trait loci important in crop performance and to recombine favorable alleles at those loci. *Oryza rufipogon*, a wild relative of rice, was crossed with three elite cultivars adapted to different growing environments in China, Korea, and Columbia, and 300 backcross lines were derived. These lines were tested in a wide range of environments and evaluated for 12 key agronomic traits. Several lines showed superior performance to the parents (Xiao et al., 1998). Subsequent marker analyses showed that, surprisingly, not all of the favorable alleles were in the adapted cultivar. *O. rufipogon* contributed alleles that were consistently associated with improvements in yield, quality, maturity, and plant height. This work, together with other examples, suggest that reservoirs of genetic diversity that reside in the wild relatives of crop species may contain numerous alleles that can provide the key to future increases in the productivity of a number of crops (Tanksley and McCouch, 1997). The great advantage of this approach is that it overrides preconceived notions of what lines or which traits might be valuable in a cross, instead allowing the genes to "speak for themselves." This is a humbling notion for scientists, and McCouch found considerable difficulty in getting breeders interested in participating in the analysis of a cross involving such an apparently useless parent as *O. rufipogon*.

The Big Problem / Opportunity

Tremendous opportunities for crop improvement are likely to arise as a result of the complete sequencing of plant genomes. This is well under way and due to be completed soon for Arabidopsis (Lin et al., 1999; Mayer et al., 1999) a rice genome sequence should not be far behind. The information that we have on plant synteny, particularly for the grasses (Gale and Devos, 1998), should ensure that the results are useful for a much wider range of plants than those sequenced. The problem is in deriving, in a timely way, knowledge of those genes that are important for crop function or for the synthesis of high-value molecules from the great mass of sequence information that is being generated.

Geoffrey Duyk (Exelixis Pharmaceuticals, San Francisco) outlined some general approaches to the problem both from a broad life science perspective and from the perspective of his own company. The innovative drive of the technology is important in that it has already reduced the cost of sequencing 10-fold since the beginning of the human genome project, and Duyk expects another 10-fold reduction in the future. Automated, high-capacity technologies and informatics have been recruited to the biological research laboratory to deal with the sequence information generated and the opportunities that it presents. His view is that this is leading to a shift in the paradigm of research as the gathering and presentation of data has become an end in itself, resulting in a dissociation of data acquisition from classic hypothesis-based research. As this process gains speed, many more genomes, particularly of important pests and pathogens, will be targeted. Exelexis is particularly focused on insects and nematodes in the hope of aiding the discovery of novel insecticides and nematocides.

John Ryals (Paradigm Genetics, Research Triangle Park, NC) discussed his company's approach to identifying new target sites to aid in the discovery of crop protection chemicals based on high-throughput DNA sequencing, high-throughput reverse genetics, and knowledge-based computer systems. Reverse genetics is based on generating transgenics overexpressing cDNAs in the sense and antisense orientation, and then analyzing their phenotype. The generation of mutants by this or any other approach (such as tagging) is likely to be one of the limiting steps. Phenotypic analysis will be automated as far as possible. The approach is to generate and digitize images of the plants, to conduct biochemical profiling via HPLC/mass spectroscopy for around 5,000 molecules of under 5,000 D, and to interrogate the database iteratively. This is likely to generate huge amounts of data that will need to be stored and accessed. A primary focus is thus on data management. It is estimated that 10 terabytes will be generated in the analysis of Arabidopsis. Such a program needs a large amount of resources focused on the ultimate objectives. Although the major multinationals were not represented at the conference, it is likely that similar resource-intensive approaches are being used by them.

An alternative company approach was described by Guy della-Cioppa (Genomics Biosource Technologies, Vacaville, CA). His company has developed a plant expression system based on RNA viruses such as the tobacco mosaic virus (Kumagai et al., 1995; Della-Cioppa and Grill, 1996; Baulcombe, 1999; Koo et al., 1999). These virus vectors are sprayed on a growing plant and rapidly multiply, giving very high expression of their inserts, which can be in either the sense or the antisense orientation. This approach allows the identification of genes whose expression leads to a marked change in phenotype in the leaves (for example, infection with a phytoene synthase insert rapidly leads to yellow plants); it may be more limited in cases in which the gene controls early development or does not function in leaves.

The major public sector initiatives were represented by Michel Caboche (Institut National de la Recherche Agronomique, Versailles, France). Unfortunately, for various reasons there were no speakers from the U.S. or other European countries. France has set up the Genoplante project. This is a 5-year scientific program, involving private and public laboratories and funding, which will share resources and data. So far, around 80 projects have been funded. The goal of the project is to find a way to co-operate with other genome projects. Its main goals are to develop expertise, infrastructure, and competitiveness in plant genome analysis. The program has two components, a generic part that focuses on model genomes (Arabidopsis and rice), and a more commercial part that will analyze the genomes of major crops (wheat, maize, oilseed rape, and sunflower) and their syntenies with the model species. The program aims to identify genes and alleles useful for molecular breeding by positional cloning and candidate gene approaches. Caboche sees the commercial approach generating industrial property rights and new biotechnology companies, but the program does not include the generation of GM crops. Results from the generic part will be published and placed in a Genoplante database. Within the Arabidopsis program, the French projects (Camilleri et al., 1998 Desprez et al., 1998 have, or aim to produce, 5,000 non-redundant expressed sequence tags, a yeast artificial chromosome library and physical map, and 50,000 T-DNA mutant lines produced by vacuum infiltration. Caboche outlined how some of these resources are being utilized within his own laboratory in the analysis of lipid metabolism in seeds.

Science and Society

It is my opinion that plant research is vital for the future of a world in which the human species has to reach a balanced and sustainable relationship with the rest of nature and its environment. We are increasing in population and this population is likely to have higher expectations in food provision and in longevity. All of this means a rapid increase in the demand for food, feed, fiber, and fuel. On the other side of the equation, we are rapidly depleting our fossil fuel reserves, and there are severe limitations on the amount of land and water available for agriculture. In addition, political problems and distribution costs suggest that we need, as far as is possible, to enable the bulk of food to be produced close to where it is needed. Eventually, if that does not happen, the hungry will move in large numbers to where there is food. The political and social consequences of a failure to meet the legitimate demands of the developing world will eventually be serious for the developed world.

There is a view that, since the West is chiefly suffering from too much rather than too little food, it can afford to put biotechnology aside and GM crops could only be used in the developing world. There are a number of dangers in this. First, there is the argument of resource efficiency: using more land, materials, and energy than we need to produce food is a misuse of resources. Second, the technology and know-how are being developed with commercial and public resources in the West, and if the technology is blocked, the flow of funds for research will inevitably diminish severely as shareholders and the public ask why money should be spent on developing this apparently unwanted technology. Third, the food supply chain is international and the decisions as to what is grown are taken by a few people in charge of the major supermarkets and commodity traders; they are more likely to support a GM-free chain for the richer customers rather than a special GM chain for the poorer customers. Finally, globalization of pressure groups is leading to universal opposition to GM crops. Thus, there is a danger that the research and development of solutions that GM crops could provide will slow or stop, despite the needs of millions of people for sufficient locally produced food. In the words of de Greef, "a food shortage caused by an empty R&D pipeline in the long-term is lethal, preventable, and immoral."

The conclusion of Braun and others is that, if scientists believe in the importance of their research and the technologies deriving from them, then they have to engage in the political process, to confront the challenges of the Green activists, to engage society in the debate, and, if necessary, take to the streets with banners. Because the outcome will most affect the young scientists, their careers, and their dependents, and because they are the ones most likely to be acceptable to society (Farmelo, 2000.), it is crucial that they are motivated to participate in the debate before it is too late. Reversing the present situation in Europe will not be easy, as supermarket chains and processed-food producers such as Gerber will need a believable case to put to the public as to why they have changed their minds on the inclusion of GM crops in their food products. However, failure to reverse the situation will jeopardize the future of plant science research and those involved in it as well as those likely to benefit from it.

Crop Improvement

The scientific goal for biotechnology and world agriculture is the improvement of the genetics of our crops. The big challenge for biological research in general, and crop improvement in particular, is how to get the most valuable knowledge the quickest from the explosion in DNA sequence information. A large-scale genomics approach is chiefly being followed in industry, but to some extent is also being tried in national and international programs. It will undoubtedly yield many benefits, but also has many limitations. It tends to be driven by data rather than hypotheses, which means that it may be linear and unprioritized (i.e. start with the first available sequence and work out the function of the genes one by one). It also faces difficulties in that many important crop traits depend on the interaction between a number of genes or may be encoded by multiple copies of the same gene. In addition, plants are renowned for their ability to compensate physiologically. Thus, generating large populations of plants in which single genes are either knocked out or overexpressed is unlikely to reveal the genetics of processes subject to these phenomena; generating multiple combinations will probably be numerically impossible. Even when the function of a gene sequence has been correctly identified, there remains the problem of finding the better or best alleles. For these reasons, I have argued elsewhere that the current thrust, which is largely based on a genocentric view, needs to be balanced by a matching emphasis on a phenocentric approach (Miflin, 2000).

A phenocentric view of crop improvement starts from the viewpoint that farmers cultivate phenotypes, i.e. crop performance is determined by the interaction of genotype and environment. While the farmer can do something to ameliorate the environment, crop improvement mainly depends on plant breeders assembling the best combination of alleles of the genes governing the key traits. The modified varieties are then widely tested to determine to which environments they are adapted. The development of marker technology, as exemplified in the work presented by McCouch, has now opened up the ability to identify many of the loci, genes, and alleles that are most important in crop improvement. It allows interactions between genes to be identified and it enables the mining of new favorable alleles from wild sources. Its crucial limitation is the ability to measure the phenotypic trait sufficiently reliably and accurately. It is also dependent on the existence of sufficient variation in the genetics of key traits in crossable species; where there is no difference between alleles at a locus, the importance of that locus in the process is unlikely to be revealed (e.g. the genes important in determining the rate of photosynthesis in wheat, in which there is little variation; Evans, 1993).

Physiology and biochemistry provide the third approach to identifying genes for crop improvement. In some cases, knowledge of biochemical pathways can lead to the identification of candidate genes, and the subsequent transgenic plants behave as expected (see the developments described by Herrera-Estrella and Potrykus). In contrast, most of the developments in trying to change carbohydrate metabolism (as described by Willmitzer) have led to the conclusion that we did not understand the biochemistry as well as we thought we did, or that compensation overrides the genetic changes. Nevertheless, much useful information has been gathered and, slowly, useful changes are being achieved. Sound physiological knowledge can also be highly useful in developing screens to probe the many populations of sequence-defined mutants that are being generated. An example of this in the past is the success of the screen devised by Somerville and Ogren (1979) in identifying many of the key steps in the pathway of photorespiration and determining their genetic control. Current technology will make this approach easier and more fruitful. All that remains is to devise the requisite "smart" screens to probe the physiological phenomena of interest.

Biotechnology & Indian Agriculture: The challenge of the Next Millenium

General Agriculture Scenario

The world has won the improtant battle in the area of food security, but the war is still on. A total of 800 million people, that is one out of every six persons, in the developing world do not have access to food. One-third of all pre-school children in the developing countries are food insecure. We, thus, have a big challenge ahead of us as we enter the next millenium.

It is true that the mass starvation that was predicted for Asia in seventies and eighties did not occur. It is only because Science was effectively put to work to raise agricultural productivity. In India, the ' *Green Revolution*' was a success due to the introduction of improved seeds, fertilizer, irrigation and plant-protection measures combined with positive policy support, liberal public funding for agricultural research and development, and dedicated work of farmers. Notwithstanding all-round achievements, the basic problems of food security, poverty, equity and sustainability, continue to be a cause of concern in India today.

The Indian Challenge.

There are several concerns about Indian agriculture that we need to address. Even in the green revolution crops, rice and wheat, the Indian average yield ranks around 50th in the world. A weak agro-based industry, the problems of weak marketing, storage, transportation, credit support, etc, haunt us. Only 0.3% of agricultural GDP is spent on agricultural R &D. The population is growing at an alarming rate of 1.8% per annum. It will be around 1.3 billion by 2020 and would require 50% additional foodgrains. A total of 200 million people are still below the poverty line and have insufficient access to food. Declining resource base, degeneration of natural resources, including soil fertility, water environment, etc, are some other concerns. Food production systems centre around a few states only. Out of 17 big states, only 7 are self-sufficient and surplus in foodgrains. There was a significant decline in per capita intake of calories and protein between 1975 and 1990.

Inspite of all these problems, there is plenty of good news. India harbors a vast diversity of basic natural resources. Tremendous bio-diversity of agriculturally important micro-organisms exists, which are not yet been harnessed. We have a variety of crops, including quality fruits, vegetables, foodgrains, and a number of plantation and commercial crops. Indian fisheries resource has spread over 8,100 km of coastline on the east and west, and provides access to marine resources and transportation.

Very large proportion of area of various food crops is under low porductivity category. It can be increased with inputs of high class science. There is a scope to cultivate sizeable portion of waste lands through soil amendment and introduction of agroforestry. Several million hectares of saturated soil can be potentially exploited from rainfed lowlands of eastern India. The India Exclusive Economic Zone (EEZ) of the marine sector, of about 2 million sq.km., has high potential harvestable yields. Only about 20% of the cultivated area is covered under tractor cultivation, great scope of making quantism quantum jump in production and productivity exists, changing consumption patterns offer scope fcr diversified agriculture. Trade liberalization provides opportunities to reach for outside markets, which were not accessible otherwise. Great gains are possible through agro-based industrialization and overall rural development.

The cultivators and farmers of India have a rich heritage of traditional agricultural wisdom on crop and livestock husbandry and fisheries; based on the principles of the conservation of natural resources and environment. India has an extensive network of extension channels for dissemination of technologies generated by research institutes. India has a strong manufacturing capability and a large network of about 20,000 manufacturers and nearly one million village artisans for indigeneous production of all types of agricultural machinery. Therefore, inspite of several disadvantages, we have a great chance to leap forward, if we enforce the right policy initiatives and implement them.

Biotechnology and India Agriculture

We need to recongnise that, just as the food reequirements of today's population of nearly 6 billion people could not have been met by the technologies of the 1940's, we cannot assume that current practices will feed the population of 8 billion expected by 2020. New approaches are needed in addition to the continued improvement of existing methods of crop and animal husbandry and food proceessing. The strategic integration of biotechnology tools into India agricultural systems can revolutionise Indian farming and usher in a new era. Genetic engineering is clearly the most revolutionary tool to impact agricultural research since the discovery of genetics by Mendel. Prior to genetic engineering, the exchange of DNA mateial was possible only between individual organisms of the same species. With the advent of genetic engineering in 1972, scientists have been able to dientify specific genes associated with desirable traits in one organism. For example, a gene from bacteria, virus or animals may be transfered into plants to produce genetically modified plants having changed characteristics. This method therefore allows mixing of the genetic material from species that cannot otherwise breed naturally.

Genetic engineering can be vital for an agrarian country like India. It can help in minimising the crop damage through disease and pest resistant varieties, reducing the use of chemicals, enhancing stress tolerance in crop plants, thus permitting productive farming on unproductive lands, etc. One could even extend the growing season of crops and minimise losses due to environmental factors on the one hand and increase the shelf life of fruits and vegetables, on the other, thus minimising losses due to food spoilage. We can expand the market vista and improve food quality. Biotechnology can also produce plants that possess healthy fats and oils, possess increased nutritive value, and create a whole range of higher value feeds. Biotechnology has even the power of producing biodegradable plastics, edible vaccines, etc.

It is only through the blending of the 'gene revolution' with our experience in the 'geen revolution' that we can reach our goal of 'evergreen revolution' and also 'nutritional revolution' The advantage of the gene revolution is that it is relatively scale neutral, benefiting big and small farmers alike, It is also environment friendly. Thus, it can be of great help to the smallest farmer with limited resources in increasing farm productivity through the availability of improved but powerful seed. It can also reduce a farmer's dependency on chemical inputs such as pesticides and fertilisers.

Modern biotechnology offers unlimited opportunity for enhancing genetic potential of crops and other commodities, management of biotic and abiotic stresses, bio-remediation and organic recycling. India is one of the main centres of agricultural biodiversity and its gene richness can greatly complement the developments in modern biotechnology. Likewise, new developments in GIS, remote sensing, and crop modelling, provide new opportunities for integrated management of natural resources. The revolution in informatics provides opportunities for sharing latest information for research and planning in a highly organised and efficient manner. With the globalisation of economy, the opportunities for value addition and post harvest management are also immense. With clever blending of technologies, the India farmer can usher in new era of confidence and performance.

Sir Francis Bacon had once said: *'It would be an unsound fancy to expect that things which have never yet been done, can be done except by methods which have never been tried'* In the new context, we have to review the new role of biotechnology. Also, it is in this context that we have to view the reapid advances that modern biotechnology is making, including the area of transgenics. Let us focus on some of the developments in this area, since considerable interest and controversy has been created around the world today in this exciting area.

Movement in Transgenics

The advent of genetic modification over the last two decades has enabled plant breeders to develop new varieties of crops at a faster rate than was possible using traditional methods, with huge potential for further beneficial development. Tobacco was the first plant to be genetically transformed\, in 1983, with cereals beginning in 1990, but it is only recently that products such as GM soya have reached the market place in Europe, and, for a variety of reasons, given rise to concern & controversy. The concern must be addressed seriously if society is to exploit the new technologies appropriately. We will address this issue rather extensively a little later. Whereas in India, we have been discussing and debating the issues, the rest of the world is galloping ahead. See

how the agro biotech product sales, includes transgenic seeds, is increasing around the world. The global area of cultivation for transgenics increased 4-5 fold from 7 million acres in 1996, to 31.5 million acres in 1997, for just 7 crops in 6 countries.

The emphasis has been mainly on traits for herbicide, insect, virus & fungal resistance, and on product quality with a limited number of genes. Out of a total 159 releases. Argentina leads the way with 43, followed by chile, Mexico, Puerto Rico and the Republic of South Africa, with 17-20 each. India is far behind with just a few releases.

There are fears relating to transgenic plants in society. These need to be addressed, not by brushing them aside, but by careful scientific analysis and educational programmes. All biosafety protocols must be followed rigidly. Let us deal with this issue more exhaustively now.

Public Concerns

The potential of biotechnology as a method to enhance agriculture productivity in the future has been accepted globally. However, because of its revolutionary nature, risk and uncertainty may be created by the process of genetic engineering and by the resulting genetically modified products. This may result from the introduction of a new unrelated DNA sequence into a recipient organisam. The introduced DNA may have some unexpected effects on the cellular processes of the recipient organism. In addition to this, introduction of antibiotic resistance genes, as selection markers, pose serious implications in public health especially in genetically modified plants that are directly used for food purposes.

Risks are also associated with the genetically modified plants that are released into the environment. The nature of interactions with other organisms of the natural ecosystems cannot be anticipated without proper scientific testing. For example, modified plants with enhanced resistance to pests or disease threaten to transfer resistance to the wild relatives, which will have implications for biodiversity and ecosystem integrity. These, and many more, doubts plague the minds of common people. Therefore, it is very clear that if the biotechnology potential has to be completely realized, it has to be done in an extremely responsible manner. At each step, proper testing and scientific data have to provided. Not only this, proper regulatory and policy mechanisms have to be put into place. It is therefore, absolutely essential, that conscious efforts be made especially by the developing countries to initiate well-defined programmes for the development and regulations of genetically modified plants. To make this possible and to benefically, ethically and sustainably reap the benefits of biotechnology, there is a pressing need for scientists, researchers, policy makers, NGOS, progressive farmers, industrialists, and representatives of the government, to come together on a common forum in order to discuss the common concerns and find solutions to them.

Will modified plants transfer their introduced genes into wild relatives growing nearby?

Will modified plants that produce new compounds disrupt the 'balance of nature' in some way?

Could the planting of a restricted number of cultivators leave crop plants more susceptible to disease?

Could the planting of a restricted number of cultivators lead to a reduction in biodiversity (of crop plants, weed species, insects and microflora in the fields in question)?

Will genetically modified plants be able to avoid the factors that regulate natural populations and threreby change the usual ' balance' between populations?

Similarly, there are concerns about the social and economic effects. For example, some of these concerns are :

How will the structure of farming (particularly in developing countries) be affected by biotechnology?

How will patent laws affect traditional breeders 'rights (e.g. the right to save seed from one year to the next)?

Will plant bredding be left increasingly in hands of a few companies, and if so, what effects might this have?

Will some countries be 'plundered' for their genetic resources?

We also have a set of ethical and moral issues that need addressing. In particular,

The consumer has the right. What will consumers be told about the new food products?

Is it acceptable to 'interfere' with nature through genetic engineering?

Do we have the right not to use all means available to improve crop plants, especially when so many people are under or malnourished?

Then there are several regulatory issues. Among these, the three key issues are:

Do current regulations give sufficient protection to farmers, consumers, those who have invested in, and those engaged in research?

Is there sufficient international legislation to ensure environmental protection?

Do current regulations compromise the competitiveness of biotechnology companies by being excessively restrictive?

The Royal society of London had appointed a group of experts to examine various aspects including the scientific evidence concerning the risk of transfer of genes from Genetically Modified (GM) crop plants to wild species and non-GM crops, the uptake of genes from GM food by the digestive system, and the current state of the regulatory system. The experts concluded that the chances of gene transfer happening are slight, provided the regulatory processes are followed, but that this must be kept under consideration.

The experts also concluded that the uptake of genes via the food chain is niot a new issue because genes (i.e. DNA) are normal constituents of the human diet. Many products from GM

plants, such as sugar prepared from GM tomato paste, are so similar that they are regarded as 'substantially equivalent' others, for example flour from GM soya, may contain a new gene or its product, although many of the purification processes involved in food prodution will destroy and DNA present in the raw material.

Some GM foods have been produced using an antibiotic resistance 'market' gene, which is a laboratory device designed to identify genetically transformed platns. The Royal Society has shown concern about the use of such genes in food products and it suggested that any further increase in the use of such markers in the human ir animal food chain would be undersirable. Gm crop plants have been produced to improve insect tolerance and virus resistance, and to include herbicide tolerance, so that transferred to non-target species of plants, and that the development of resistance by target pests is minimised, the regulatory authorities must be assurd that any negative effects would be no greater than those resulting from conventional procedures; and that any long term effects on the environment and ecology would be clsoely monitored, with statutory restrictions in place to control marketing; and also that best practice advice was adopted by growers.

The reliance on a case by case approach to the legislation, may result in lack of analysis of the overall impact of the technology on agriculture and the environment, and of the long term effects of gmoS. aLTHOUGH MECHANISMS ARE ALREADY IN place to regulate many individual aspects of GM technology, there is no means for looking at GM technology as a whole. As independent, overarching, regulatory body is needed to span departmental responsibilties, monitor the enforcement of existing or future regulations, and strengthen the guidelines to growers of GM crops, such as those specifying the isolation distances between Gm and non-GM crops. The body should also review and monitor the membership of advisory committees and regulatory bodies.

The Indian Effort in R & D

Indian has set up a large network of reasearch institutions and agricultural universities located in dirrerent regions and states of the county.

Main Thrust of Activities of Crop Biotechnology

Agriculture in India accounts for nearly 65% of the country's employment, 26% of the total GDP and nearly 20% of total export earning and supplier of raw material to major industries. Agriculture is not only the backbone of Indian economy and food security but also a way of life, a tradition and anchor of overall livelihood opportunity for about 700 million of our one billion population. Agriculture, therefore, is and will continue to be central to all strategies for planned socio-economic development of the country. Despite major advances in agriculture and strong growth in food production in the latter part of the twentieth century, hunger and famine are still common in many parts of the country. The application of biotechnology techniques within the agriculture sector can potentially improve food security by raising crop tolerance to adverse weather and soil conditions, by enhancing adaptability of crops to different climates and by improving yields, pest resistance and nutrition, particularly of staple food crops

Genetic modification of crops is one such method allowing individual characters (gene, factor or trait) to be transferred into crop plants. With the advent of genetic modification through genetic engineering in early eighties, the natural barrier of only intra-specific exchange of characters was removed and scientists were able to identify and transfer specific genes associated with desirable traits from one organism to the organism of other species that otherwise cannot breed naturally. With these techniques genes from varied class of organism like bacteria, viruses or even animal may be transferred into plants to develop genetically modified plants having exclusively changed characteristics controlled by the specific gene. This gives scientists/breeders a broader access to desirable traits from any living organism and its possibility of transferring it with much faster rate and greater precision.

The Department has successfully built-up a strong base for taking advantage of this innovative science. Continuous efforts during 9th Plan have added strength to this mission. The Department has made concerted efforts for supporting various programmes in this area since 1990. Presently, there are three types of ongoing research activities: R&D Projects on priority crops, Multi-institutional projects. The Department has supported plant molecular biology programmes in the six institutions; Madurai Kamraj University, Madurai; Tamil Nadu Agricultural University, Coimbatore; Osmania University, Hyderabad; Bose Institute, Calcutta; NBRI, Lucknow and Delhi University South Campus, New Delhi.

The mandate of these programmes relate to major problems of priority crops, development of transgenics, quality improvement and basic research in the area of plant molecular biology. There are a number of multi-institutional projects e.g. in wheat programme was initiated for characterization of quality traits; in mustard-development of molecular methods for hybrid seed production; development of transgenic cotton, rice, mungbean and tomato for resistance to various biotic stresses; salinity and dehydration stress tolerance in rice; cloning of responsive genes and their promoters and development of wheat varieties with dial resistance to leaf and stripe rust using molecular marker technology.

The Department took the initiative to join the International Rice Genome Sequencing Programme (IRGSP) in June 2000 with a commitment to sequence 10 Mb of Chromosome 11 in five years. The programme is running ahead of time and India has fulfilled the international obligation in two and a half years time, till date a total sequence data of 15.80 Mb has been deposited.

Achievements

Under the wheat network project different approaches were used to develop molecular markers for tagging genes with quality traits. However, PCR based approaches detected adequate polymorphism. STS and STMS markers linked to three traits (pre-harvest sprouting, protein content and grain size) were identified; also ISSR, RAPD and STMS markers linked to protein content and grain size were identified.

In another project on development of molecular methods using barnase and barstar genes for hybrid seed production, in mustard transformation of *B. juncea* cv Varuna with constructs has been initiated. A number of transgenics have been developed with different barstar constructs and plants were selected *in vitro*, on herbicide barstar. Constructs have been designed in which influence of 35 S promoter is limited and transgenics with barnase gene showing male sterile flowers have beenobtained. Insect resistant transgenic rice developed through introduction of a modified endotoxin gene of Bt. Further bioassay of putative transgenics and their progeny for expression of Bt gene is being carried out.

Molecular constructs have been made both for insect and viral resistance. Transgenic technology has been standardized and found to be most amenable to cotton and rice. Transgenic system of indica rice and wheat has been developed. Rice has been transformed with codA/COR47, AtHSP100 and PDC gene, while wheat was transformed with HVA1 to confer stress resistance. All the transgenics have been characterized at the molecular level for gene integration.

For developing *Vigna mungo* plants with yellow mosaic virus resistance, coat protein gene in sense orientation and replicas gene in both sense and antisense orientation have been cloned. The binary vector has been mobilized into *A. tumifaciens*. Transformation of *V. mungo* and selection of transformed plants is in progress.

Studies have been conducted to transform IR-50 rice plants with P5CS gene to make it more tolerant to salinity. The assessment of transformed plants harboring this gene in single copy and double copies showed promising results.

For incorporation of synthetic Bt Cry 1A(b) gene into vegetable *Brassicas*, the presence of transgene was checked via PCR. Southern hybridization was also done. Individual transgenics were found to carry 1-4 copies of Cry 1 A(b) gene. Northern and Western analysis confirmed the expression of transgenes.

For tagging genes for LF resistance in rice, two RAPD markers linked to a major gene of leaf folder resistance have been identified. Based on these markers SCAR primers have been developed.

For tagging genes for brown plant hopper resistance in rice, several RAPD and SSR markers for BPH resistance have been identified. Also conversion of a dominant RAPD marker AJ9 to a Co-dominant STS marker done and this marker is used in the selection of genotypes for BPH resistance.

Two salt inducible genes, PC kinase and proline serine rich protein encoding genes (PCSDR) were isolated from cDNA library of *Porteresia coarctata*. Both the genes have been characterised at *nucleotide* level.

Interesting leads obtained was on salt tolerant inositol synthase gene from halophytic rice, *Porteresia coarctata*.

The reproducible protocol for somatic embryogenesis and direct shoot regeneration has been established for three popular sugarcane varieties viz. Co.J.64, Co.J.83 and Co.J.86.

Leads Obtained

After standardization of transgenic technology, now emphasis is on development of large number of transgenics in cotton, rice, mungbean and tomato followed by diversification, molecular characterization and field evaluation. The international obligation of Rice genome sequencing data was fulfilled by the end of December 2002. Under the collaborative programme, transgenic *Brassica* is being tested for the male sterility, herbicide resistance and *Alternaria* blight in the field conditions. Transgenic system in indica rice and wheat have been developed. Rice has been transformed with codA/COR47, AtHSP100 and PDC gene while wheat transformed with HVA1 to confer stress resistance. All the transgenics have been characterized at the molecular level for gene integration. Studies have been conducted to transform indica rice plants with P5CS gene to make it more tolerant to salinity. The assessment of transformed plants harboring this gene as a single copy showed promising results and further efforts are on to pyramid one more gene for higher tolerance to salinity.

Technologies / Products

The application of biotechnology techniques in the agriculture sector can potentially improve food and nutritional security by raising crops tolerant to adverse weather and soil conditions, by enhancing adaptability of crops to different climates and by improving yields, pest resistance and nutrition, particularly of staple food crops. Over the past decade, the application of biotechnology to the problems in world agriculture has yielded significant productivity gains to the producers. Various programmes of the Department have been directed in this direction. With advancements in GM technologies, these benefits are expected to increase in Indian scenario.

New Initiatives for the 10th Plan

The rice genome-sequencing programme with present commitment of gap filling and annotation shall be given high priority.

A collaborative network programme on functional genomics of rice would also be considered. It has been decided to support one collaborative multi-institutional project each in the area of 'strategic research' and another on 'impact research and delivery'. The following two projects are under consideration (i) gene expression profiling during flower and seed development and functional validation of identified genes and (ii) identification and functional validation of genes related to yield and biotic stresses.

It is intended to start a Wheat Biotechnology network with an aim to sequence wheat genome and mapping of important traits.

A collaborative programme on the use of biotechnology for improvement of semi-arid crops like Sorghum and Chickpea in association with ICRISAT and National Research Laboratories would be initiated.

A network project on the use of molecular marker technology and genetic engineering approach in wheat quality breeding shall be launched on the basis of leads obtained from previous project.

As a follow-up action of Indo-US workshop on agricultural biotechnology, it is proposed to initiate a major multi-institutional programme on abiotic stress tolerance in rice, chickpea and wheat.

Like previous years, the application of molecular biology tools in crop breeding with special reference to improvement of quality, protection from biotic/abiotic stresses will receive high priority.

Exploitation of hybrid vigour by developing and using genetically engineered male sterility in selected crops.

A programme on engineering the biosynthetic pathway of starch in rice and wheat would be considered.

Development of marker assisted selection techniques for use in breeding for tolerance to drought, submergence and salinity in selected crops.

The ongoing efforts using molecular biology tools for protection against biotic stresses will receive high priority. The list of target crops shall be increased with the perfection of technology.

For combating malnutrition, efforts would be made to transform high yielding varieties of staple food crops with enhanced protein contents, development of rice varieties with p- carotene and transformation of rice and wheat for high iron content.

Engineered male sterility for extended hybrid technology to more crops shall be exploited.

Major genes representing Quantitative Trait Loci for tolerance to abiotic stresses (drought, moisture, salinity, heat, heavy metals and toxic substances), resistance to biotic stresses (viruses, bacteria, fungi, insects, nematodes), plant architecture (height, leaf area and thickness, root profile), flowering time, yield and mineral nutrition would be the main targets of study and developing transgenic crops..

Fine mapping of genome regions harboring useful genes would be done by taking advantage of available sequence data of genomes to tag genes for marker-assisted precision breeding.

8

Agroecology, Sustainable Farming & Agrobiotechnology: Myths & Risks

Introduction to Agroecology

Agroecology is a scientific discipline that uses ecological theory to study, design, manage and evaluate agricultural systems that are productive but also resource conserving. Agroecological research considers interactions of all important biophysical, technical and socioeconomic components of farming systems and regards these systems as the fundamental units of study, where mineral cycles, energy transformations, biological processes and socioeconomic relationships are analyzed as a whole in an interdisciplinary fashion.

Agroecology is concerned with the maintenance of a productive agriculture that sustains yields and optimizes the use of local resources while minimizing the negative environmental and socio-economic impacts of modern technologies. In industrial countries, modern agriculture with its yield maximizing high-input technologies generates environmental and health problems that often do not serve the needs of producers and consumers. In developing countries, in addition to promoting environmental degradation, modern agricultural technologies have bypassed the circumstances and socio-economic needs of large numbers of resource-poor farmers.

The contemporary challenges of agriculture have evolved from the merely technical to also include social, cultural, economic and particularly environmental concerns. Agricultural production issues cannot be considered separately from environmental issues. In this light, a new technological and development approach is needed to provide for the agricultural needs of present and future generations without depleting our natural resource base. The agroecological approach does just this because it is more sensitive to the complexities of local agriculture, and has a broad performance criteria which includes properties of ecological sustainability, food security, economic viability, resource conservation and social equity, as well as increased production.

To put agroecological technologies into practice requires technological innovations, agriculture policy changes, socio-economic changes, but mostly a deeper understanding of the complex long-term interactions among resources, people and their environment. To attain this understanding agriculture must be conceived of as an ecological system as well as a human dominated socio-economic system. A new interdisciplinary framework to integrate the biophysical sciences, ecology and other social sciences is indispensable. Agroecology provides a framework by applying ecological theory to the management of agroecosystems according to specific resource and socio-economic realities, and by providing a methodology to make the required interdisciplinary connections.

Agroecology of the Abonera System: A Case Study

The *abonera* system is similar to an improved, short-term fallow. After the maize crop is harvested, the field is abandoned to the spontaneous growth of velvetbean vegetation. The function of the velvetbean fallow is to help maintain and build up soil productivity for the benefit of the following maize crop. No direct economic benefits are realized from the fallow vegetation *per se*.

The thick mulch layer that velvetbean leaves on the soil year-round is a distinctive feature. In this respect, the *abonera* system is more like natural ecosystems, such as tropical forests with litter layers, than rotational fallow cropping systems (see Budelman 1988). One of the main effects of the velvetbean-mulch layer is improved mineral nutrition in the maize crop. The mulch layer also has favourable cumulative effects on soil fertility and reduces soil erosion, making the *abonera* system a viable, productive long-term option for continuous cultivation of hillsides.

In this study, two central aspects of the *abonera* system are examined to show how the system works and how it is related to the climate, soil, and other aspects of the surrounding environment. First, this study provides baseline information about annual nutrient cycling in the *abonera* system, with a strong emphasis on N dynamics. The main aspects considered here include quantification of organic inputs, pace and timing of N accumulation in the legume, mulch decomposition, and the related accumulation of inorganic nitrogen (N_i) in the soil profile. Uptake of N by the maize crop is also examined. The evidence for this analysis comes mainly from measurements of velvetbean biomass made in farmers' fields at slashing time and from monitoring N_i in the soil profile throughout the maize cycle in a subsample of these fields.

N dynamics has been studied in numerous related agroecosystems (Huntington et al. 1985; Ladd and Amato 1985; Yost et al. 1985; Glover and Beer 1986; Pichot et al. 1987; IRRI 1988; Yost and Evans 1988; Sanchez et al. 1989; van der Heide and Hairiah 1989; Palm and Sanchez 1990; Smyth et al. 1991; Kang and Mulongoy 1992; Haggar and Beer 1993). What is much less common, however, is empirical evidence from developing countries on the long-run effects of cropping systems, and this is due to the high costs of maintaining experiments (Pieri 1989; Swift et al. 1991; Sanchez

1994; Steiner 1995). This study analyzes the agroecological sustainability of the *abonera* system by comparing farmers' fields managed continuously under the *abonera* system for periods of 1–15 years. We analyze soil samples collected in chronosequences of the velvetbean–maize rotation in four villages in northern Honduras. The set of properties examined includes levels of soil organic matter (SOM) in the upper soil profile, soil acidity and exchangeable bases, P content, infiltration, bulk density, and macroporosity.

A chronosequence approach allows inferences to be made about the evolution over time of a system. To obtain the data for this, a comparative survey is conducted at a given point in time on a set of fields that are supposed to represent successive historical states of the system. The use of this space-for-time substitution scheme is a common practice in ecology studies (Pickett 1988) and in soil-genesis studies. It is much less common, however, in cropping-system studies (Staley et al. 1988; Feller et al. 1991; Kleinman 1995), probably because the proper conditions for the use of such schemes are rare. This approach also entails many assumptions that make the analysis more vulnerable to failure. Nevertheless, it was the only alternative at hand that does not require at least a decade of observations before an experimental database is available for the formulation of recommendations. We believed, moreover, that a chronosequence scheme was particularly pertinent in northern Honduras, where numerous contiguous fields are managed in similar ways for various periods.

Nitrogen Cycling

This section analyzes annual trends in aboveground velvetbean biomass and N dynamics and their relations to the availability of N_i in the soil profile during the maize cycle. We analyze management options for meeting the N requirements of maize by determining the effects of limited additions of fertilizer-N or fertilizer-P on maize production.

Main Components of the Aboveground Biomass

Aboveground biomass in the velvetbean–maize rotation includes several key components, whose nutrient content or relative contributions to total biomass vary with the phases of the rotation. As a first simplification, the aboveground biomass can be divided into "live" and "dead" fractions. The live fraction com-prises either growing velvetbean (from June to December) or growing maize and its accompanying weeds (between December and May). The biomass content of this fraction varies widely during the year, reflecting the various phases of the velvetbean–maize rotation. In all cases, however, no biomass other than maize ears is ever removed from the velvetbean fields: maize stover is left in place, and velvetbean is never grazed or harvested as forage or grain.

The dead fraction consists of a dead mulch or litter layer *sensu strictu*, which completely covers the soil surface year-round. Components of the litter include a dynamic mixture of decaying velvetbean parts, decaying weeds slashed by farmers during the maize cycle or suffocated by velvetbean during the major season, and rotting maize stover. The biomass content in this layer is always high, contributing consistently more than 50% of the total aboveground biomass found in a velvetbean field at any given time. Biomass content reaches its highest levels after slashing of velvetbean and again following the incorporation of maize stover in the litter.

Velvetbean Biomass and Nutrient Xontent at Slashing

Slashing of the live velvetbean crop constitutes the pivotal moment in the velvetbean rotation. We will now turn our attention to two fundamental aspects of slashing, namely, the quantity of biomass present at that time and its composition.

Total Biomass Content

For the four villages sampled in December 1993 (Las Mangas, Piedras Amarillas, Rio Cuero, and San Francisco de Saco), the average levels of total aboveground biomass fell within a relatively narrow range of 10.7–12.4 t ha^{-1}, on a DM basis, although differences among sites were statistically significant. Similarly, the year-to-year variability within a site was moderate, although biomass was significantly lower in December 1992 than in the two following cycles. The largest differences occurred among fields within the same year and site. For example, in San Francisco de Saco in 1992, the individual-field minimum was 6.1 t ha^{-1}, whereas the maximum exceeded 15 t ha^{-1}. The within-field variability was low on average (not statistically significant), although in a few cases differences of several tonnes per hectare were found between sampling plots within a single field.

The overall stability of the biomass production (coefficient of variation of less than 15%) probably stems from a combination of factors. First, total biomass includes a strong semipermanent litter component, which is only partly influenced by seasonal fluctuations in climate and plant growth. Also, the length of the velvetbean cycle (8 months minimum) probably allows the velvetbean–weed stand to compensate over the growing season for any stress that would temporarily reduce growth. To go beyond total biomass and grasp its actual composition, we subdivided the live fraction into three subfractions — green material (velvetbean leaves and fine stems), velvetbean pods, and velvetbean vines (partly lignified stems) — reflecting the morphological and functional differences of this fraction's components.

Characteristics of the Biomass Fractions

Across sites, the levels of N and the C–N ratios were fairly consistent among the various components of the biomass. The pods had the highest level of N (about 3% on average); the vines had the lowest (less than 2%), which translates into a C–N ratio of greater than 23. The litter fraction had relatively high but variable levels of N within site: about 2.65%, on average, with consequently low C–N ratios of 16–18. The litter had a ^{13}C value close to -26 (^{13} is a measure of the relative abundance of ^{13}C in the plant tissues [Mariotti 1991]). The N and ^{13}C values indicate that the litter fraction at slashing received a greater contribution from the velvetbean, a C3 plant, than from maize stover (a C4 plant with a ^{13}C value of close to -13) or from C4 grass weeds (*R. cochinchinensis* in the case of San Francisco de Saco), which predominate in numerous velvetbean fields across northern Honduras. By contrast, when velvetbean did not reestablish itself properly during the 1994 summer cycle, the biomass at the next slashing had a much higher proportion of weeds, yielding a lower N content (less than 2%) and lower ^{13}C values for the litter fraction (-15 to -20).

Nitrogen content

Total N content in the aboveground biomass averaged across all four sites. As was the case for total biomass, total N content was similar across sites and reached almost 300 kg ha^{-1} on average. Again, the major source of variability was among fields: at San Francisco de Saco, for example, content dropped to as little as 100 kg ha^{-1} in one field and, conversely, reached almost 500 kg ha^{-1} in another. As N content was fairly similar across all sites, total N mainly depended on biomass levels, rather than on the N content of the various fractions.

Consequently, almost 60% of the N present at slashing was found in the litter, rather than in the live subfractions. The pod subfraction accounted for a low percentage of total N on average, but not at Las Mangas, where it accounted for almost 25% of the total N, in keeping with the high proportion of pod biomass at this site. Because velvetbean seeds will eventually germinate, most of this pod N is not expected to be available for subsequent recycling via decomposition.

Nutrient Content

Although N is the nutrient of primary interest in this discussion, the accumulation of other key nutrients in the velvetbean biomass was quite significant. Considerable variability occurred across sites, but the velvetbean "complex" accumulated large quantities of Ca (140 kg ha^{-1} on average, 70% of it in the litter) and K (100 kg ha^{-1}, 82% in the live subfractions). Even P (15–20 kg ha^{-1}, 45% in the litter) was found at levels adequate for the requirements of a moderately high yielding maize crop.

Seasonal Behaviour of the Velvetbean Cover

We now examine in more detail how a velvetbean crop accumulates DM and nutrients in the first place, before releasing both on decomposition.

Biomass accumulation during the rainy seaso The two main phases of the velvetbean cycle are the vegetative, from February and March (velvetbean reseeding) to early October, and the reproductive, from October to December, at which time velvetbean starts to die naturally, even without slashing. The vegetative phase spans the dry season and the first half of the rainy season, whereas the reproductive phase takes place during the peak of the rainy season.

TABLE 1

Nutrients (other than N) in the aboveground biomass of velvetbean at four sites at slashing time, northern Honduras, December 1993

Nutrient	*Average*[a]	*Minimum*	*Maximum*
Total P (kg ha^{-1})	20±7	14	28
% of total P in litter	45±14	31	58
Total K (kg ha^{-1})	100±24	82	114
% of total K in litter	18±9	11	27
Total Ca (kg ha^{-1})	140±37	111	159
% of total Ca in litter	70±10	62	78
Total Mg (kg ha^{-1})	26±7	22	32
% of total Mg in litter	56±12	45	67

[a]Each cell represents the average over all fields sampled at four sites, followed by its standard deviation. Sample sizes were 32, 29, 21, and 19 for San Francisco de Saco, Las Mangas, Rio Cuero, and Piedras Amarillas, respectively.

After it reseeds itself, between February and March, the velvetbean grows relatively slowly because of the shade provided by a fully developed maize crop. Also, it has to withstand either farmers' pruning operations in wet years or extremely dry and hot conditions if the winter cycle is drier than usual. Weeds left uncontrolled by farmers may also compete heavily with the young velvetbean plants for light, nutrients, and water. Usually not until after maize harvest and the return of rains (by end of May to early June) do conditions become favourable for rapid velvetbean growth, leading to full canopy closure within a few weeks. By mid-summer, a typical field has a relatively uniform, dense velvetbean stand. By then, maize stover has been pulled down and incorporated in the litter by aggressively growing velvetbean vines, which had been using the stalks as support. Weeds have usually been reduced to a marginal presence by that time, as velvetbean gradually outcompetes most of the weeds present at the end of the maize cycle. Velvetbean starts flowering in early to mid-October, apparently in response to shorter days (it is still unclear how strictly photoperiodic velvetbean is). At this point, a typical velvetbean field in northern Honduras has accumulated about 10 t DM ha^{-1}, with close to 40% of the DM in the live velvetbean fraction and slightly more than 60% in the litter layer.

But biomass accumulation does not stop at flowering. During the 1993 cycle, total biomass increased from 10 t ha^{-1} in mid-October (early flowering) to 12 t ha^{-1} a month later and to 14 t ha^{-1} after another 3–4 weeks. DM accumulation seemed to affect the litter layer more than the live fraction. Between mid-October and mid-November, the biomass contained in the live fraction remained roughly stable, around 3.5 t ha^{-1}, or less than 30% of the total biomass, only to increase by about 1 t ha^{-1} by slashing time in mid-December, probably as a result of accumulation of DM in the pods themselves. On the other hand, litter biomass increased sharply (from 6.4 t ha^{-1} in mid-October to 8.7 t ha^{-1} in mid-November and to almost 10 t ha^{-1} at slashing; data not shown), which may indicate that although velvetbean does not die massively until it is slashed, it starts decaying before or soon after flowering, by shedding leaves and stopping maintenance of its extensive vine network.

As expected, the overall accumulation of N by the velvetbean complex matched closely the trends observed for total biomass. Total N for all fractions increased from 289 kg ha^{-1} in mid-October to 334 kg ha^{-1} in mid-November to 367 kg ha^{-1} by slashing time, giving an overall rate of N accumulation of about 1.3 kg ha^{-1} d^{-1}. Again, the situation differed markedly for each fraction: whereas the live fraction apparently accumulated no net N during the 2-month period, the litter fraction gained 89 kg ha^{-1}. This gain, together with a decrease in the C–N ratio, is consistent with the net transfer of biomass from the live fraction to the litter fraction via leaf-shedding, hypothesized earlier. This has important implications for nutrient release and recycling, which start significantly before velvetbean slashing and follow closely the addition of this fresh, N-rich material to the litter layer, where abundant rainfall favours its rapid decomposition. Further evidence that this is the case is found in the N_i levels in the soil profile.

Mulch Decomposition during the Dry Season

After farmers have slashed the velvetbean stand, decomposition is the major process affecting the litter layer. The table shows apparent, rather than actual, rates of decomposition because periodic samplings of unconfined material made it impossible to separate out the decomposition of the litter *per se* from litter renewal via the addition of fresh weed biomass during weed-control operations.

Litter biomass appeared to drop markedly at first, from 12.6 t ha^{-1} at slashing to 8.6 t ha^{-1} in early March, corresponding to a loss of about 45 kg ha^{-1} d^{-1}. From March to the end of May, however, the total biomass present in the litter layer seemed to increase, reaching 10.6 t ha^{-1}. Although it may be an artifact stemming from approximate sampling procedures, this increase may also reflect the impact of weed-control practices during the February–March period. That slashing the weeds or drying them out with paraquat actually contributes new biomass to the litter layer is an interpretation supported by an analysis of *in situ* labeling, provided by the ^{13}C values (Balesdent et al. 1988).

According to these calculations, the original velvetbean litter decomposed relatively quickly from December to early March, losing 43% by weight during this period, and much more slowly afterwards, losing only an additional 6%. Weeds controlled by farmers contributed significant quantities of new litter during the same period: by the end of May, the weeds seemed to represent almost 40% of the litter found (4 t ha^{-1} out of a total litter of 10.6 t ha^{-1}). This table does not include the biomass of live weeds, which can range anywhere between 0.5 and 4 t DM ha^{-1}. The situation in terms of N was similar to the one for biomass: total N (for the entire litter) dropped sharply between December and March, from 316 kg ha^{-1} to 198 kg ha^{-1}, and increased again to 235 kg ha^{-1} by late May, paralleling the apparent biomass increase. Using the same sort of calculations on natural abundance mentioned previously, we estimated the quantities of N remaining in the original velvetbean fraction. N content dropped from 316 kg ha^{-1} to 176 kg ha^{-1} in early March, to 171 kg ha^{-1} in late May. About 140 kg ha^{-1} of N seemed to have been released by the litter on average in the first 80 d following slashing; and less than 5 kg ha^{-1}, in the following 80 d (however, the variability associated with both estimates was huge).

It is probable that these crude figures, obtained in a very dry cycle, represent lower than average estimates of the N released during a typical (that is, fairly wet) winter cycle, especially after March when there are usually at least a few significant rains. However, the behaviour of the velvetbean cover during two phases (fast then slow release) seems consistent with observations of the decomposition of green manures elsewhere (Bouldin 1988).

Nitrogen Dynamics in the Soil–maize System

The main objective of this section is to gain some understanding of the relation between N supplied by the decaying litter–SOM and the N demand and uptake from the maize crop (in terms of quantities and synchronization). This issue is critical to overall system productivity, as the N released by the velvetbean is of little direct value to the maize crop unless its availability is in synchrony with the demand for it from the maize.

Temporal Patterns of Inorganic Nitrogen

The general temporal patterns exhibited by N_i (sum of NO_3-N and NH_4-N [kg ha^{-1}] over the entire 0- to 60-cm soil profile) over the 1992/93 and 1993/94 maize cycles for a number of neighbouring, well-established velvetbean fields (each with 5–14 years of continuous use of the velvetbean–maize rotation). Several features are:

- All fields displayed a relatively homogeneous behaviour in when N_i was highest and in how fast its level changed. The similarity of pattern within and between years illustrates both the homogeneity of management across fields and the strong influence of environmental factors and conditions (other than the soil) in shaping N mineralization processes.

- Each year, N_i reached a marked peak at about 30 d after slashing, followed by a rapid decrease over the next 3 or 4 weeks. The maximum observed levels of N_i were close to 100 kg ha^{-1} for both years (the maximum observed was 115 kg ha^{-1}; the minimum, 70 kg ha^{-1}). They never dropped below 30–50 kg ha^{-1} N_i, even during the period of maximum maize uptake.
- As maize is planted, immediately after slashing, a satisfactory synchronization can be obtained between N released by the decaying velvetbean mulch and that taken up by the maize crop. The sharp decrease observed in the levels of available N_i between days 30 and 80 (1992/93) or 90 (1993/94) — during which 60–80 kg ha^{-1} of N_i disappeared — coincided with periods of intense crop uptake. In addition to maize, weeds are also likely to have benefited from the high levels of available N_i, especially in the first few weeks after slashing, when maize grows slowly.
- A sizable pool of N_i (40 kg ha^{-1} or more) was available in the profile even when maize was not growing. This is especially evident from the 1993/94 data, which cover a longer time span (from October to June). In particular, the relatively high levels of N_i (around 60–70 kg ha^{-1}) found in the profile in October–November (that is, well before slashing) may indicate that active decomposition is taking place in the litter layer–SOM complex during the heaviest rains, when velvetbean is still growing. This trend is consistent with the increase in litter biomass caused by leaf-shedding, mentioned earlier.

Distribution of Inorganic Nitrogen by Horizon

As the season progressed (that is, as maize went from emergence to flowering), the profile was gradually depleted of its N_i at all depths. Toward the end of the maize cycle, availability of N_i tended to increase again, especially in the top horizon, which became the main contributor to total N_i. Even in 1993/94, a dry year, its share reached around 50% of the total N found in the profile.

The dynamics affecting the first horizon over time and the observed difference between the upper and lower horizons probably reflect the influence of maize–weed uptake, as roots preferentially deplete the N_i of the superficial horizons. In terms of concentrations, the order showed a strong gradient — horizon 1 > horizon 2 > horizon 3 — a situation typical of a no-tillage system with a litter layer. The decreases in concentrations over time in the various horizons may be due to decreasing availability of substrate and moisture, which are both key factors in the decomposition process.

Nitrogen Released by the Decomposing Litter

The decomposing litter alone appeared to release about 100 kg ha^{-1} in the first 80 d after slashing (this value differs from the one reported because it was calculated only for the fields monitored for N_i). How much of this N found its way into the soil solution remains a matter of speculation, as the N may have been volatilized (Costa et al. 1990) or immobilized by the fauna inhabiting the litter or simply intercepted by plant roots at the litter–soil interface before entering the soil profile (Schlather 1997).

Nitrogen Released by the Mineralization of Soil Organic Nitrogen

An estimate of the N mineralized from SOM during the maize cycle can be derived by estimating mineralization rates for the organic N stored in the soil profile. This was a humid, tropical climate and, on average, the moisture content of the various horizons remained favourable to mineralization from December until at least early March; therefore, mineralization rates in the 0- to 10-cm horizon may have reached 1–1.5% of the N present for this 3-month period (based on a 4–6% annual rate). If the estimated contributions of the different horizons are added together, N mineralized in the 0- to 60-cm profile from the pool of soil organic N alone could have contributed about 50–75 kg N_i ha^{-1} between early December and early March, thus adding significantly (as much as 50%) to the levels of N released by the decaying litter.

Nitrogen Budgets

By summing for the main sources or sinks of N over the maize cycle, one can calculate approximate N budgets at various stages in the maize cycle. The terms taken into consideration in this analysis include the following:

- The aboveground litter, with dynamics discussed earlier;
- The maize crop, with the assumption that the N-uptake curve over time was of the form $a/(1 + b \exp -ct)$, where a, b, and c are fitting constants and t is time (Hunt 1982);
- Weeds, which constitute another important but variable factor, both in N uptake (especially when weed control is deficient) and in N recycling (in effect, because of the way weeds are controlled [slashed manually or desiccated with paraquat], most of the N the weeds take up can be expected to be recycled later during the growing season [Lambert and Arnason 1989]); and
- The N_i found in the 0- to 60-cm soil profile (its level reflects the mineralization of both the litter and the soil organic N).

The N unaccounted for (obtained as the difference between the total N at a given stage and the total N at slashing) is a measure of how well the above representation (and measurement) of the various N pools holds (Legg and Meisinger 1982). Overall,

TABLE 2

Estimated N budgets at several stages in the maize cycle, San Francisco de Saco, northern Honduras, winter 1993/94.

Stage	*N (kg ha⁻¹)*					
	Litter[a]	*Maize*[b]	*Weed*[c]	*Soil* N_i[d]	*Total N*	*N unaccounted for*[e]
Field A						
Slashing	335	0	0	90	425	–
30 DAP	284	7	30	96	417	-8
50 DAP	255	45	10	54	363	-61
70 DAP	225	90	20	45	380	-45
Harvest	272	100	41	45	458	33
Field B						
Slashing	299	0	0	80	379	—
30 DAP	264	7	30	75	376	-4
50 DAP	244	47	10	45	346	-34
70 DAP	224	94	20	39	377	-2
Harvest	217	105	46	42	411	31
Field C						
Slashing	238	0	0	68	306	—
30 DAP	234	6	30	48	318	12
50 DAP	231	41	10	46	329	22
70 DAP	229	83	20	32	364	57
Harvest	204	92	33	38	368	61
Field D						
Slashing	312	0	0	67	379	—
30 DAP	284	7	30	53	374	-5
50 DAP	266	47	10	39	362	-17
70 DAP	248	94	15	26	384	5
Harvest	253	105	17	33	408	29

Note: Each value represents the mean of three replications. DAP, days after planting.

[a]Total N found in aboveground biomass (live fraction + dead fraction), which is measured.

[b]Total N in aboveground maize biomass, estimated using a logit function, except for harvest, which is measured.

[c]Total N in weed biomass, estimated, except for harvest, which is measured.

[d]Inorganic N as measured in 0- to 60-cm soil profile.

[e]Calculated as total N_s — total $N_{slashing}$, where *s* is the given stage in the maize cycle.

these budgets appear to offer reasonable approximations, as the N unaccounted for represents relatively small amounts (a small percentage of the total N at slashing, with maximum deviations of 10–20%). In other words, the various components taken into consideration in calculating these budgets (litter, crop, weeds, and N_i in the profile) seem to represent the most important ones, with little room for losses via leaching or other forms of N immobilization.

Nitrogen Leaching

In an N-rich environment subject to abundant rainfall, leaching is the probable fate of N released by the decomposing litter–soil, and the calculations obtained above for N budgets are compatible with leaching's playing a role. Following an approach developed by Jones (1975), we estimated apparent leaching rates by calculating for each field the average depth of the N_i for the various sampling dates and regressing this against the cumulative amount of rainfall. The results of these calculations showed that in both cycles (1992/93 and 1993/94), no downward movement of N_i was apparent.

This indirect evidence, combined with indications from the N budgets and from N_i monitoring, indicates that leaching is probably not a significant source of N loss in the velvetbean system, at least not during the maize cycle. But it may be a result of the gradual release of N from the decomposing litter and the temporary trapping of N in weeds, which may play a significant role in protecting N against leaching early in the maize growing season. This is when rains are still frequent and heavy and the maize crop is still unable to take up much N. Further evidence is required, however, to assess leaching losses more fully.

Nitrogen Stored or Otherwise Immobilized

Besides plant uptake, there are two likely sinks for N released by the decomposition of litter and SOM: either the soil–litter biota or the SOM itself. No data are available on the former, although it may be expected that microbial biomass, in particular, should demonstrate a strong seasonality in response to the increased availability of the substrate produced by slashing. In all likelihood, the turnover of this N is relatively fast (Duxbury et al. 1989) and hence the net release of some of it is possible, even within the maize cycle, still in time for subsequent plant uptake.

Evidence of the long-term role of SOM as a sink is given by the generally positive trend observed for soil organic N values in the 0- to 10-cm horizon: from about 0.2% in check fields where no velvetbean has ever been grown to more than 0.3% for old velvetbean fields. The gradual increase observed over the years corresponds to an overall storage of about 50–80 kg N ha^{-1} $year^{-1}$, on average. How much of this yearly storage occurs during the maize cycle remains unclear.

Recycling of Nitrogen Versus N_2 Fixation

It is usually assumed that much of the N in legume–cereal rotations comes from biological fixation. We made no direct measurements of this during the study, but the data gathered on N cycling allow us to make some indirect estimates.

Disregarding losses via leaching and volatilization and assuming stable levels of the microbial pools of N across years, we considered two mechanisms by which annual N cycles are kept open: nutrient removal via harvest (grain only) and long-term storage in SOM (at least until the system reaches near equilibrium).

Each term representing about 50–80 kg ha^{-1} $year^{-1}$, a total of 100–160 kg N ha^{-1} $year^{-1}$, must be obtained from an external source to make up for the loss or storage. Some N may enter via rainfall or nonsymbiotic fixation (perhaps, 20–30 kg ha^{-1} $year^{-1}$ [Wetselaar and Ganry 1982]), but most probably the bulk of it is provided through symbiotic N_2 fixation by the velvetbean crop itself. Until direct, *in situ* measurements are made, it appears reasonable to conclude that a velvetbean crop may fix anywhere between 70 and 130 kg N ha^{-1} per cycle.

Conversely, as much as 200–300 kg N ha^{-1}, or about two-thirds of the total N, is simply recycled through the system every year. The velvetbean crop (and to a much lesser extent, the weeds) appears to be a primary candidate for scavenging any available N, because of the large biomass it accumulates, the amount of time it has to accomplish this task (almost 6 months, as the sole or major crop), and the conditions highly favourable for mineralization and litter decomposition found during this wet period. In addition, one may expect velvetbean to rely as much as possible on the ample supply of N_i in the environment, rather than incurring the high energy costs of fixing all the N it needs (Giller and Wilson 1991).

Maize Response to Nitrogen

In this section, we examine whether the "natural" supply of N from the velvetbean biomass meets the N requirements of a maize crop and how sensitive this is to the levels of N present in the soil–plant system. The two main N inputs for a growing maize crop in the velvetbean system are the N provided by the decomposing velvetbean-biomass fractions constituting the litter, on one hand, and the soil organic N, on the other — both of which release N gradually upon mineralization. Fertilizer-N can constitute a third source for those farmers willing and able to invest in such a costly input.

Fertilizer can at best add flexibility to the management of the velvetbean system, as velvetbean can almost instantaneously increase N availability for plant uptake over that spontaneously released by the soil–litter organic complex. It may thus contribute to higher yields, for which ample N is needed at critical stages in crop growth, irrespective of its source (organic or chemical). Nevertheless, adding even more N may constitute a wasteful use of precious cash and labour, as well as an undesirable contribution to N leaching in the environment, given the large quantities of N present in the velvetbean system.

Maize Response to the Nitrogen Accumulated by Velvetbean at Slashing

Data on the response of maize to N accumulated by velvetbean at slashing comprise N content of the litter at slashing, maize-ear-leaf N content at silking, and maize yields. Clearly, factors other than N supply influence the relationships among these variables. However, a number of trends are apparent.

First, highly significant differences ($P < 0.001$) between sites appeared in maize-ear-leaf concentrations. Similarly, highly significant differences ($P < 0.001$) were found between years at the same site. In 1993, maize appeared to have achieved a better nutritional status vis-à-vis N (average % N = 2.81) than in 1994 (average % N = 2.41), although the levels of potential N supply, measured by N content of the biomass, were similar for both years. A weak tendency for yields to increase in response to corresponding increases in maize-ear-leaf concentrations was evident.

Using an interpretative approach based on envelope curves (Siband and Wey 1994), we inferred that N supply might possibly have been limiting if less than 70 kg ha^{-1} was present in the live fraction (which translates into levels of total biomass at slashing of less than 8 t ha^{-1}). Above this threshold, maize appeared to have no clear response to N content, as maize yields seemed to reach a plateau around 5–5.5 t ha^{-1}. Most of the velvetbean fields had N levels of more than 70 kg ha^{-1}, the threshold mentioned earlier, but also had yields much below the alleged plateau (2–3 t ha^{-1}, compared with 5 t ha^{-1}). This may indicate the likelihood of limiting factors other than potential N supply. Overall, the lack of a clear trend.

Relationship between maize yields and N content of (A) the biomass and (B) the maize ear leaf, northern Honduras, 1992/93 and 1993/94 cycles. Note: Each point represents the average of three (in 1993/94) or four (in 1992/93) replications. Slashing refers to the approximate date for manual cutting of the *Mucuna*, which is left to decompose on the soil surface. Planting, flowering, and harvest refer to approximate dates for the maize cycle. M, San Francisco de Saco, 1993; San Francisco de Saco, 1994; X, Las Mangas, 1994;, Piedras Amarillas, 1994; + Rio Cuero, 1994.

Maize Response to Adding Nitrogen Fertilizer

How maize in the velvetbean system responded to a single dose of urea-N applied at the rate of 50 kg ha^{-1} is the focus of this section. The fertilizer was applied, not as a replacement for the N from the decomposing velvetbean litter, but as a complement to this organic source. Whereas the previous results came exclusively from survey data, the fertilizer analysis was carried out by establishing a total of 14 simple, replicated on-farm experiments.

Variability of the Response to Nitrogen Fertilizer

Variability in response among and also within fields (from one replication to another) was important. In practical terms, it reflects the fact that certain fields showed no response to N (or a very weak one), whereas others responded sharply (yield increases of about 1 t ha^{-1} were obtained in a few cases). Three of the 10 fields analyzed in 1993/94 showed a statistically significant response to N, whereas 7 showed no significant response when taken individually.

The average response by site (across fields) to N was markedly different between years. In 1992/93 (a relatively wet winter cycle), maize showed no significant response to N, and average yields showed no increase with the application of fertilizer-N (P = 0.21).

In 1993/94, however (a rather dry cycle), overall response to N was highly significant ($P < 0.01$), with an average yield increase over the case of no fertilizer reaching about 0.6 t ha^{-1}. Interestingly, similar behaviour was observed for maize nutritional status vis-à-vis N: visual symptoms of N deficiencies, almost absent in 1993, were widespread in 1994, and ear-leaf N concentrations at silking were significantly lower in 1994 than in 1993 and were also significantly affected by the application of fertilizer (2.31% without N versus 2.50% with N in 1993/94; $P < 0.0001$).

The fact that fertilizer is advantageous in a relatively dry year may appear contrary to conventional wisdom. But it is hardly surprising if one considers that drier conditions slow down decomposition of the surface mulch without affecting the water availability in the soil profile too drastically. In effect, the rainfall pattern and the fairly deep soils characteristic of the hillsides of northern Honduras allow a winter maize crop to start growing with a soil profile holding up to 300 mm of stored water, which serves as a buffer against potentially extended dry spells. Moreover, a mulched profile can efficiently conserve this stored water (Steiner 1994).

Possible Causes of the Differential Response to Nitrogen

The marked variability detected in maize response to N (which was confirmed further in a series of 15 fertilizer-N trials conducted in San Francisco de Saco and Rio Cuero in the 1994/95 cycle [Barreto, personal communication, 1996[3]]) raises the question of why a given velvetbean field or block responds to fertilizer-N. We conducted a qualitative assessment of this issue by analyzing the values taken for a series of possible factors in maize response to N, using three incremental classes of response to N obtained at the repetition level: nil, weak, and strong.

This analysis yielded some statistically significant differences ($P<=0.01$) between repetitions belonging to class 0 (nil response), on the one hand, and repetitions belonging to classes 1 and 2 (weak and strong responses, respectively), on the other hand. In particular, repetitions from class 0 presented the highest maize yield levels, highest maize N status, and highest soil organic N levels. Whereas the N content of the velvetbean production was equivalent among classes, the repetitions that showed no response to fertilizer-N had been on average cropped for more time in the velvetbean system than the ones responding markedly to N (10 years for the former versus 7 years for the latter), indicating that response to N is more likely in younger velvetbean fields than in older ones. All other variables, whether reflecting environmental conditions (such as rainfall received) or management (such as timing of weed control), presented similar levels for all classes.

Farmers' Practical Knowledge About Nutrient Dynamics and Soil Fertility

Evidence from the N_i monitoring showed that, by deliberately planting maize almost immediately after slashing, farmers were placing their crop in a good position to take advantage of the flush of N_i entering the profile, which in effect brought about an almost ideal synchronization of the crop uptake and N environment. Considerations other than nutrient availability constrain the choice of a slashing–planting date. Slashing has to be delayed until viable seeds are produced. However, the interval during which it is desirable to slash also depends on the need to avoid possible drought or weed competition. Given these constraints, farmers seem content to cleverly pattern management to the ecology of velvetbean without trying to modify its basic parameters.

Many farmers don't use fertilizer-N when growing maize in rotation with velvetbean, although generally it is locally available and they know about its use. When asked about their rationale for not using fertilizer-N, many stated very clearly that a major reason (even before considering costs) for not applying urea to their maize in rotation with velvetbean was that their fields didn't need it. In effect, they consider velvetbean a green manure produced *in situ*, replacing external fertilizer. As noted previously, it is no coincidence that the local name for velvetbean is the fertilizer bean (*frijol de abono*). Conversely, farmers readily recognize that maize planted without velvetbean responds markedly to fertilizer-N applications.

In sum, although farmers in northern Honduras have never formally experimented with velvetbean decomposition patterns and fertilization, many of these farmers have for the most part already assimilated the bulk of the practical knowledge about its use in their management practices. We turn now to the analysis of long-term trends, using a chronosequence approach at four sites distributed in northern Honduras.

Long-term Changes in Soil Properties

Overall changes in content of soil organic matter in the 0- to 10-cm horizon

On a regional basis, the tendencies observed in San Francisco de Saco were not entirely confirmed at other sites. At Las Mangas, no changes in C or N contents appear to have occurred, but at Rio Cuero, the changes seem quite dramatic, even after only 7 years in the velvetbean rotation. Also, the levels of C or N found in the check plots vary significantly across sites, undoubtedly reflecting differences in edaphoclimatic conditions and perhaps in agricultural history at the village level (the fact that San Francisco de Saco exhibited the lowest levels of both elements appears consistent with both its lower elevation and its being the oldest human settlement of the sites studied). In no cases did the older velvetbean plots have less C or N than the check plots. Stated conservatively, then, the velvetbean rotation appeared to allow at least the conservation of the initial stocks of C and N, despite continuous annual cultivation. Relative changes in total C and N contents of the 0- to 10-cm horizon in the velvetbean system, San Francisco de Saco, northern Honduras. Note: Each point represents one observation plot; dotted lines represent least-standard regressions.

Relative changes in C content in the 0- to 10-cm horizon in the velvetbean system at four sites, northern Honduras. Note: Each bar represents the average for a given age class and site, topped by its standard error. Sn Fco, San Francisco de Saco; Mangas, Las Mangas; Cuero, Rio Cuero; Piedras, Piedras Amarillas.

Distribution of Soil Organic Matter in the Upper Soil Profile

Soil samples were collected in 2.5-cm increments in the upper 15 cm of the soil profile to verify whether the accumulation of C was mainly affecting the top centimetres of the soil profile, as expected in any no-tillage cropping system (Follett and Peterson 1988; Barreto 1989; Dalal et al. 1991). Changes in C content were significant in the first 5 cm of the soil profile, especially in the 0- to 2.5-cm layer, in which the relative increase was about 50% over a decade (from 3% to 4.5%). For the 2.5- to 5-cm layer, over the same period, the increase was 40%, with a peak value of 2.8%. Conversely, no apparent increases were detected for layers between 7.5 and 15 cm: all plots had a uniform C content, regardless of their age.

Furthermore, the regression approach to the analysis of these two horizons yielded in both cases a quadratic term that was statistically significant. This may indicate that a leveling off of the C accumulation is taking place after about 9–10 years of rotation.

Changes in the distribution of organic C by 2.5-cm increments in the first 15 cm of soil profile under the influence of the velvetbean–maize rotation, San Francisco de Saco, northern Honduras. Note: Vertical bars represent standard errors. Horizon 1, 0–2.5 cm; horizon 2, 2.5–5 cm; horizon 3, 5–7.5 cm; horizon 4, 7.5–10 cm; horizon 5, 10–12.5 cm; horizon 6, 12.5–15 cm.

Changes in Chemical Properties

A sensible hypothesis for long-term evolution of the soil profile is that acidification is likely to take place in a wet tropical environment such as that of northern Honduras, as potential imbalances between an ample supply of N from the velvetbean biomass and moderate uptake by the maize crop might rapidly induce significant leaching of N, along with its accompanying cations (Bouldin 1989; Cahn et al. 1993). However, soil-test results for both pH and exchangeable Ca and Al did not present any evidence to support this hypothesis. In San Francisco de Saco, after 15 years of continuous use of the velvetbean rotation, pH appeared to have remained fairly constant, around 6.0, throughout the entire soil profile (up to 60 cm), with perhaps even a slight (not significant) tendency to increase over time. Likewise, levels of exchangeable Ca and Mg appeared to have increased over time at all depths at three out of four sites, to reach levels close to 15 cmol(+) kg^{-1}. At these levels of pH and exchangeable bases, one expects to find no free Al ions, a condition confirmed by the soil-test results. Similar but less clear results were obtained at other sites.

Changes in pH and Al and Ca contents over time in the velvetbean–maize rotation, San Francisco de Saco, northern Honduras. Note: Each point represents the age-class average for each horizon.

The increase in exchangeable bases has also been observed in other mulched systems (Lal 1989; Hulugalle et al. 1990); it may be attributed to the large yearly additions of Ca and Mg via the velvetbean biomass — reaching more than 150 kg ha^{-1} year^{-1} in the case of Ca, or the equivalent of more than 0.6 cmol(+) kg^{-1} of Ca if applied to the 0- to 10-cm horizon. How and from what source the velvetbean crop mobilizes this Ca remain a matter of speculation. In sum, the absence of soil acidification is consistent with the previous observation that although the availability of potentially leachable N_i was high during the maize cycle, there was little evidence of actual N_i leaching.

Phosphorus

Together with N, P is a very common limiting factor in crop production throughout the tropics. In systems that include a legume to supply N, a shortage of P frequently becomes a major obstacle to sustained yields (Schlather 1997). Hence, maintaining an adequate supply of available P over time is a critical concern in the velvetbean system.

Across sites, a sizable variability occurs in both trends detected and quantities of available P. Overall, however, a conservative view is that P availability seems to remain fairly stable in the velvetbean system despite yearly exports (via harvest) amounting to about 15–20 kg ha^{-1} year^{-1}, an observation well in accord with the velvetbean system's lack of response to fertilizer-P. As is the case for all other nutrients, decomposition of the velvetbean biomass is undoubtedly a major source of available P: yearly additions of P via the aboveground biomass reach about 15–20 kg ha^{-1}.

Changes in Physical Properties

We analyzed historical trends for one site only (San Francisco de Saco) and for a limited subsample of the fields analyzed for chemical properties or SOM. As might be expected for soil physical properties measured on few plots and at a microscale, within- and between-field variability was fairly high. This contributes to the uncertainty of the analysis (see Horowitz 1995).

Erosion

Soil erosion was not actually measured in this study. Given the overwhelming importance of soil erosion in hillside farming, however, some general, qualitative comments are in order.

The characteristic signs of erosion at the field scale were virtually absent, even in the oldest velvetbean fields (with more than 15 years of continuous cultivation). Gullies or rills were seldom observed, except in very localized areas, where rill erosion seemed more a result of marginal management errors than of anything else. Also, the upper horizon presented none of the enrichment in coarse material typical of areas with significant surface runoff (Foster et al. 1985). The chemical analyses demonstrated that no depletion of nutrients was occurring over time and that the upper profile was actively accumulating organic matter, as well as being comparatively richer in nutrients than the underlying horizons. These observations suggest that little erosion was occurring.

On a larger scale, small creeks collecting water at the bottoms of the slopes in the velvetbean rotation were very clear, even during or after intense rains, whereas high sediment loads could be observed at the bottoms of neighbouring, unmulched slopes.

Other evidence is more difficult to interpret. As discussed in this study, as many as 40% of farmers surveyed reported that the velvetbean system might induce localized landslides in areas with very steep slopes (more than 60—70% slope). Our discussions with farmers confirmed that such landslides occur once in a while (not every year) during the peak of the rainy season (anytime between September and November), under very heavy rainfall.

A possible explanation for these landslides includes a combination of the heavy weight of the wet velvetbean biomass; a loosening of the upper soil profile, as a result of the shallow rooting habits of the velvetbean plant; and a state of supersaturation of the soil resulting from increased infiltration, inducing a lower shear strength and higher overburden weight (Van Es, personal communication, 1995[4]). Some farmers also indicated that landslides might be explained by a lack of deep rooting or anchoring, caused by the substitution of the traditional bush-fallow rotation for one with a fairly shallow-rooted species, such as velvetbean. Furthermore, if left unpruned, velvetbean is quite able to eradicate the few trees left in place by farmers.

None of these possible explanations is completely convincing. The landscape in the mountains of northern Honduras is geologically very young, not fully stabilized. Hence, mass redistribution continues to take place spontaneously in many areas, and landscapes with abrupt slopes are most likely to be affected by this gravity-driven redistribution process (whether such landscapes should be cultivated on a large scale is definitely a relevant question). Also, one can argue that if hundreds of millimetres of water pour on any landform in a few hours or days, something dramatic is likely to happen; the actual role of the velvetbean cover in causing a landslide is probably much less significant than that of the sheer masses of water rushing their way downhill. This may explain why landslides, when they take place, affect lands under all kinds of use, from virgin forest to pastures, to fields without velvetbean, and have no obvious preference for any one category of land use. When 400–700 mm of rain fell in a 15-h period (31 October 1993), it caused countless landslides in the hillsides.

The issue seemed important enough to be addressed in a general survey of the velvetbean system conducted in the summer of 1994. Farmers were specifically asked about the occurrence of landslides before and after introducing the velvetbean rotation in their fields. Out of 34 fields that had suffered landslides (from a total of 44 fields included in the survey), 21 (62%) had had similar problems before velvetbean was ever introduced. Furthermore, only one-third of the farmers attributed the landslides to velvetbean use. Their perceptions varied strongly from village to village: in Piedras Amarillas, where landslides are common, farmers blamed velvetbean for making things worse, whereas in San Francisco de Saco, where landslides are rare, most experienced velvetbean users vehemently opposed this view.

In sum, it is fair to say that globally, the velvetbean system is extremely efficient in preventing erosion, as it creates and maintains year-round a thick mulch that protects the entire soil surface from the direct impact of rain. The evidence with regard to the landslide issue remains inconclusive, and further assessment is needed.

Bulk Density and Macroporosity

Bulk density and macroporosity tell us a lot about the health of a soil profile, as they are indicators (not necessarily unambiguous or direct) of important synthetic properties such as soil structure and, to a lesser degree, root and faunal activities.

Bulk density — Average bulk-density values at the field level tended to decrease in the soil profile. In horizon 1, bulk density dropped sharply from an initial 1.36 Mg m^{-3} to about 1.20 Mg m^{-3} in old velvetbean fields (the regression of bulk density on years in velvetbean was highly significant, with $P < 0.01$). For the lower horizons, the drop was smaller: from 1.41 Mg m^{-3} to about 1.33 Mg m^{-3} in horizon 2 (not significant) and from 1.45 Mg m^{-3} to 1.37 Mg m^{-3} in horizon 3 ($P < 0.04$).

These trends are consistent with what one expects the velvetbean roots (most of which are very shallow) to achieve without tillage, as well as with the increase in SOM. Interestingly, they also reflect the increased looseness or softness of the upper profile, induced by velvetbean use, that many farmers reported in qualitative terms.

TABLE 3

Changes in bulk density of the 0- to 10-, 10- to 20-cm, and 40- to 50- cm horizons in the *Mucuna*–maize rotation, San Francisco de Saco, northern Honduras

Years with Mucuna	*Bulk density (Mg m⁻³)*		
	0–10 cm	*10–20 cm*	*40–50 cm*
No *Mucuna*	1.36±0.145	1.41±0.065	NA
1–2	1.32±0.066	1.32±0.072	1.45±0.022
4–7	1.20±0.075	1.32±0.124	1.40±0.067
8–11	1.28±0.083	1.37±0.066	1.42±0.076
>12	1.20±0.091	1.33±0.064	1.37±0.070

Note: Each value represents the average for a given age class and depth, followed by its standard deviation. NA, not available.

Macroporosity — As indicated by the bulk-density figures, total porosity increased over time, especially in the 0- to 10-cm horizon. We hypothesized that, regardless of what happened to the total porosity, shifts in the distribution of pores of different sizes may occur as a result of velvetbean use. We examined this hypothesis by quantifying the porosity associated with the biggest pore-size classes (pores ranging in diameter from 15 to more than 395 μm).

In the 0- to 10-cm horizon, we found that immediately following the introduction of velvetbean rotation, the porosity associated with both pores larger than 15 μm and those larger than 133 μm increased (from less than 8% of the soil volume to about 10% and from about 3% to about 5%, respectively). The porosity remained virtually stable in subsequent years. For the 10- to 20-cm horizon, porosity was essentially identical for all fields and was quite high in all but one case.

The clear increase in porosity detected for the largest pore sizes may be a slight exaggeration because of unavoidable imperfections in the construction of the chronosequence. What is most striking, however, is that the velvetbean rotation appears to allow the soil to maintain an extensive array of large pores, without any tendency for degradation of this favourable pore architecture.

Changes in macroporosity (pores of >15 μm diameter) in the 0- to 10-cm and 10- to 20-cm horizons under the influence of the velvetbean–maize rotation, San Francisco de Saco, northern Honduras.

Infiltration

The steady-state infiltration rate — a parameter directly related to an intrinsic profile property (Hillel 1982) — increased markedly with time under the influence of the velvetbean–maize rotation, although variability within and between fields was quite high.

Using a multiple-regression approach to accommodate the effect of a number of variables under the specific conditions of the infiltration measurements, we found that, on average, steady-state infiltration rates increased by 2–3 mm h^{-1} for each year of the velvetbean rotation. Over 15 years, this led to an increase of more than 30 mm h^{-1}, roughly double the initial rates measured in no- or young-velvetbean situations. Conversely, runoff rates (observed under a simulated rainfall intensity of 100 mm h^{-1}) decreased by about 2 mm h^{-1} $year^{-1}$ on average, from 72 mm h^{-1} in no-velvetbean fields to a low of 26 mm h^{-1} in old velvetbean fields.

Changes in steady-state infiltration rates under the influence of the velvetbean–maize rotation, San Francisco de Saco, northern Honduras. Note: SSI = $-0.2738a^2 + 7.097a = 29.77$ ($R^2 = 0.757$), where SSI is steady-state infiltration; and a is age (years in *Mucuna*).

Water Balance in the Velvetbean System

Profile recharge and water-holding capacity are affected by higher infiltration rates and porosity and make more water available to the maize crop, supporting such pivotal biological activities as decomposition and mineralization. This may be particularly important during winter cycles with a marked dry season and may be even more so in drier environments than in northern Honduras, with water balance becoming a critical parameter in crop production.

One should be careful not to underestimate the consequences of decreased runoff rates and intact porosity for erosion. As runoff is reduced, the erosive action of rainfall is also drastically decreased. Because the velvetbean system provides a year-round, full cover for the soil surface, it prevents even occasional high levels of runoff from translating directly into high erosion or soil-structure degradation (such as surface sealing [Bielders and Baveye 1995]). The runoff flows on top of or in the mulch layer, rather than over a bare surface.

Long-term Changes in Crop Productivity

Does maintenance or build-up of soil fertility (as reported earlier for a number of components of global soil fertility) translate into increased crop (maize or velvetbean) productivity? We addressed this by looking at trends in maize yields with the velvetbean rotation. To assess the validity of our analytical findings, we also asked farmers to evaluate the long-term changes in the soil fertility of their fields.

Changes in Average Maize Yields and Yield Components

The trends for average maize yields at each site for different age classes. Four main conclusions can be drawn from these data:

- Average maize yields varied markedly by site (from a low of less than 2 t ha^{1} in Rio Cuero to a high of 4.4 t ha^{-1} in Las Mangas in 1993). The ranking seems to reflect at least partially the difference in overall soil fertility (pH and availability of exchangeable bases in particular).
- Maize yields with velvetbean were almost double those obtained for crops without velvetbean (Rio Cuero was an exception, with an increase of only 40%, but the study at this site included only one check plot).
- Once the velvetbean rotation is well established (more than 3 years), yields seemed to remain fairly constant. In fact, there was no apparent tendency for yields to decline. Maize yields have a tendency to be more stable in the older velvetbean fields: the standard deviation across sites dropped from 1.5 t ha^{-1} in younger velvetbean fields in 1992/93 to 0.73 t ha^{-1} in fields with velvetbean for 8 or more years; and from 1.0 t ha^{-1} to 0.7 t ha^{-1} in 1993/94.
- Maize-yield components, such as the number of ears per plant and number of kernels per ear, provide an additional basis for analyzing the effects of soil fertility on crop productivity, as they are good indicators of favourable growing conditions before flowering (Fleury et al. 1982; Navarro Garza 1984; Fleury 1991). In our case, these components demonstrated a significant improvement with velvetbean in the 1992/93 cycle (the situation was not as clear-cut in the 1993/94 cycle).

The role of another yield component, plant density, is more complex, as it partly depends on management decisions. A drop in plant density seems to explain the apparent drop in yield in fields with velvetbean for 5–7 years (1992/93 and 1993/94 cycles), as well as the old velvetbean fields' failure to outyield the medium-term ones in 1992/93. Plant densities were lower in check plots than in fields planted to velvetbean, probably as a consequence of farmers' deliberate adaptation to an increase they perceived in soil fertility with the use of velvetbean.

Years in velvetbean and plant stand were significant predictors of yield levels in a multiple-regression approach for all sites and years, with the exception of Rio Cuero. Using the slope of the equations obtained in the various cases, we concluded that, on average, every additional year in velvetbean results in an extra 50–170 kg ha^{-1} of maize, whereas every additional 5000 plants harvested results in an increase of 250–500 kg ha^{-1} in maize.

Especially noteworthy from a qualitative viewpoint is perhaps the greater stability apparently provided by the velvetbean rotation in the face of adverse climatic conditions. There is less risk of low yields with the velvetbean rotation than with other cropping systems. The probability of achieving a yield level less than or equal to 1000 kg ha^{-1} is about 70% in the first season in the bush-fallow system; about 62% in the second season in the bush-fallow system; and only 40% in the *abonera* system (a second-season maize cycle). The distribution for second-season maize in an *abonera* system lies everywhere to the right of the other two distributions (Sain and Buckles 1997). The lower risk of losses in second-season maize production for farmers who use the *abonera* system significantly influences their evaluation of the benefits of this system.

Field observations also point to the same trend. During the drier than usual 1993/94 cycle, many maize fields suffered from drought stress. In some villages, yields from fields without velvetbean dropped to very low levels (less than 1 t ha^{-1} or even complete crop failure), whereas nearby fields planted to velvetbean around the same time were faring reasonably well (2 t ha^{-1} or more). This implies improved access to water for maize in velvetbean fields, which can be ascribed to a combination of reduced evaporation and better infiltration, as indicated earlier.

Farmers' Evaluations of Long-term Changes

Farmers using the velvetbean rotation were asked to compare yields they could reasonably hope to get before using velvetbean with those after it was firmly established in their fields. In the most extreme cases, farmers reported that velvetbean helped to triple their yields or even to reclaim fields they considered too poor to produce a maize crop. By contrast, some farmers reported no changes at all or only very minor yield increases. Averaged across sites, the reported yield gain reached 70%, from about 1.5 t ha^{-1} to 2.6 t ha^{-1}.

It is also interesting to note that an overwhelming majority of farmers (43 out of 46) thought that the soil quality of their fields had improved with the introduction of velvetbean (soil was judged "better" or "much better" by equal proportions of the farmers). Many of the farmers claimed to deliberately use higher plant densities in velvetbean fields because they felt that the velvetbean made the soil more productive. No farmers reported degradation resulting from velvetbean use. In collective interviews conducted at the village level, farmers were explicitly asked to consider any negative behaviour or characteristic affecting velvetbean fields with the passage of time, but they could identify none. Not even the risk of landslides seemed to increase in older fields. Another solid indication of the improved quality is the higher sale prices and rents fetched by land in velvetbean than fetched by average farm land; the difference can be as much 70% (Buckles et al. 1992; Humphries, in press).

Altogether, velvetbean farmers were extremely satisfied with the agronomic results of velvetbean use. A final proof is that none of the farmers seemed to have ever abandoned a velvetbean field for agronomic reasons.

The Rottboellia Puzzle

One long-term negative trend in productivity was reported by farmers in San Francisco de Saco, Atlántida. Beginning in the middle to late 1980s, an obnoxious weed, *R. cochinchinensis*, began invading *Mucuna* fields. In this section, we give a brief explanation of why this happened and examine its impacts on the *abonera* system.

Rottboellia may have been introduced to northern Honduras with food aid (mainly rice) brought in after the devastating typhoon Fifi. The weed spread rapidly to the communities lying very close to the main road along the coast, including San Francisco de Saco (Munguia 1992). *Rottboellia* found an ideal environment in the *abonera*. Most other weeds, especially broadleaf weeds, are eradicated by the *Mucuna* crop. But nutrients, light, and water were readily available after the slashing of the *Mucuna* field and before full establishment of the maize crop. Farmers' reliance on natural reseeding of *Mucuna* also allows gaps to form in the *Mucuna* stand, which are quickly invaded by *Rottboellia*. Elimination of the weed after it is firmly established is extremely difficult because its reproductive cycle is very rapid and it produces massive amounts of seed with great longevity (Bridgemohan et al. 1991). These factors favoured the rapid establishment of *Rottboellia* in *abonera* fields in some communities.

Farmers in San Francisco de Saco indicated that the cost of weed control increases markedly in fields invaded by *Rottboellia*, and maize yields decrease by as much as 0.5 t ha^{-1} on average. Efforts to prevent its spread and eradicate it if it is firmly established seem so futile that farmers in this community have come to accept *Rottboellia* as unavoidable in the *abonera* system. The farmers argued that the added cost and inconvenience are a small price to pay for what is otherwise a very productive and beneficial maize-production system. Our own research on the role of weeds in nutrient recycling suggests that fast-growing *Rottboellia* may capture nutrients early in the season that would otherwise be lost. These nutrients would be released later when the weed is cut — a process documented for other weed species in cropping systems (Lambert and Arnason 1989). The puzzle is whether *Rottboellia* will be a significant threat to long-term crop productivity.

Synthesis: How Does the Velvetbean System Work?

Perhaps it is worth restating that the challenge of continuous cultivation without long-term fallowing (especially on hillsides) is the rapid decline of soil fertility from erosion or nutrient mining. As soil fertility declines and noxious weeds start competing strongly with crops, crop yields drop, farming becomes both very tedious and unprofitable, and fields are abandoned. This is the point of reference needed to judge the achievements of the velvetbean system.

The most important conclusion from our analysis is that the velvetbean system shows us a working example of how to sustainably exploit the properties and dynamics of a natural ecosystem for the benefit of commercial maize production (Gliessman et al. 1981). At the heart of the velvetbean system lies the velvetbean crop: the crop acts alternately as a major collector (when growing) or supplier (when decomposing) of nutrients, so its natural seasonal dynamics dictate the major features of the velvetbean system.

The multilayered structure of the velvetbean system explains its overall dynamics. At any given time, at least two distinct layers (or compartments) are functioning in concert. One layer is the growing, live biomass (in effect, a crop–weed mixture). It actively accumulates nutrients under the driving force of photosynthesis. Depending on the phase of the cycle, the crop is either in velvetbean or in maize and its associated weeds. Whereas the function of maize in the system is relatively straightforward, the function of the growing velvetbean is more complex, ranging from controlling existing weeds to recycling or fixing N to shielding the underlying litter or soil from direct exposure to the heavy rains.

The other layer of the system is a semipermanent dead-litter layer, which, together with the first few centimetres of soil, serves as a major provider of nutrients for the growing biomass. The litter is from the natural or farmer-induced decay of velvetbean, maize, or weed biomass. Its continuous presence and multiform activity throughout the year make it a prime regulator of nutrient fluxes, acting both as a substrate for decomposition and as an almost ideal habitat for decomposing flora and fauna that thrive in a microenvironment, as well as protecting that environment from brutal variations in temperature and moisture. Alone or in association with the live biomass, velvetbean fulfils several other key functions, such as weed and erosion control (the latter, by cushioning the impact of water drops and favouring infiltration).

The litter layer is maintained by two opposite processes: litter formation and litter decomposition. Part of litter formation is farmers' management of maize, weeds, and velvetbean, which codetermines, with environmental conditions regulating plant growth, the levels of addition to the litter, as well as its timing. Each of the three main components added to the litter also has distinct initial properties vis-à-vis decomposition. For example, velvetbean typically has a high N content, low C–N ratio, and very leafy, easily decomposable material, but the opposite is true of maize stover. Although the processes of litter decomposition may fluctuate markedly in response to periodic additions of fresh material, they seem only moderately affected by management. They are largely under the influence of environmental factors, such as moisture and temperature (Jenkinson 1981). These two factors continually interact to modify the microclimate of the litter layer and its ability to decompose.

Nutrient Cycles

With the *abonera* system, farmers derive substantial nutritional benefits for their maize crop. They let the velvetbean accumulate *in situ* the biomass and nutrients needed for the succeeding second-season maize cycle. These nutrients are gradually released through the decomposition of the velvetbean mulch, which is created by slashing and deliberately maintained by the farmers' decision not to burn it. The velvetbean's N_2-fixing and recycling abilities prevent significant nutrient losses to the environment and practically eliminate the need for costly and impractical (on a hillside) use of external fertilizer, without compromising yield levels.

Symbiotic fixation of N_2, estimated at 80–150 kg N ha^{-1} $year^{-1}$, is crucial in balancing the N budget. This newly fixed N helps counterbalance both the export of N via the maize harvest (typically in the range of 50–80 kg ha^{-1} $year^{-1}$) and the storage of N in the SOM, which may reach 50 kg ha^{-1} $year^{-1}$, at least in the first 10 years of the rotation (in the years after this, the soil seems to achieve a certain equilibrium). It remains unclear whether fixation is actually more important in the initial years following velvetbean introduction and then drops to maintenance levels after a significant pool of recyclable N is established.

The velvetbean system appears to recycle large quantities of nutrients throughout the year via the velvetbean and the weeds. For a dry cycle like 1993/94, more than 200 kg N ha^{-1} was recycled. This magnitude is comparable to that of a number of natural forestry or agroforestry ecosystems (Vitousek and Sanford 1986).

As in natural ecosystems, the losses of N in the *abonera* system unrelated to crop exports (that is, leaching, volatilization) seem relatively limited (at least under the conditions where our data were gathered). However, losses from leaching may be higher in very wet winter cycles, when decomposition is probably fairly active. We also cannot rule out the possibility of significant losses through volatilization after slashing.

Long-term Trends in Soil Fertility

Continuous use of the velvetbean system has a number of cumulative effects on the soil profile. Here again, the net impact is a balance between the results of processes that tend to deplete the stocks of nutrients (such as repeated exports via crop harvest) and decrease the soil fertility and the results of processes that tend to replenish the stocks (such as N_2 fixation) and increase the soil fertility.

SOM accumulates at or very near the soil surface as a result of the humification of the litter. Water infiltration increases markedly, as does total porosity. Despite a theoretical possibility that N imbalances lead to soil acidification, measurements indicated that soil pH remained stable, although exchangeable bases tended to increase throughout the soil profile. At the same time, most other nutrients remained at stable levels, at least in their available forms, despite yearly exports. These observations confirm that the velvetbean system is a relatively efficient nutrient-cycling system, as was mentioned earlier. Although soil biological life was not measured, it appeared to prosper, as indicated by the proliferation of earthworms, insects, and fungi at the litter–soil interface. The quasi absence of serious pests or soil-borne pathogens also points to a healthy functioning of the soil profile.

Finally, although erosion was not measured directly in this study, it was visibly only a marginal occurrence in most velvetbean fields because of the permanent protective cover provided by the velvetbean biomass (live or dead), day after day, year after year. Velvetbean use may, however, contribute to localized landslides at certain sites with excessive rainfall and slope. But no clear strategy against these occurrences seems possible, as even native forest is not immune to landslides.

In short, the long-term indicators we were able to examine gave positive or at least satisfactory results. Perhaps the clearest indication of success, from a user's perspective, was that maize yields in old velvetbean fields were actually as high as or even higher than in new ones and, on average, about double those from check plots not planted to velvetbean.

Conclusions

A maize crop benefits in many ways from the environment and general dynamics of a well-established *abonera* system. First, the system seems fairly stable, allowing respectable yield levels (usually 2–4 t ha^{-1}) every year. In particular, the system appears to prevent or at least greatly diminish drought stress because the mulch layer helps conserve water in the soil profile (Steiner 1994). With enough water around, nutrients (N, P, K, Ca, Mg, etc.) are made readily available, in good synchronization with major crop uptake. In addition, the *abonera* system creates a relatively trouble-free environment for maize because most weeds (with the notable exception of *R. cochinchinensis*) have a hard time flourishing in this system, either because velvetbean physically prevents them from germinating and emerging or from surviving very long during the velvetbean cycle or because a shallow rooting of weeds in the litter layer–soil interface makes them easier to control.

The *abonera* system has, however, a number of minor constraints. One is the tight coupling of maize planting with slashing. Until alternatives are found (for example, the introduction of velvetbean or maize germplasm of different maturity classes), the planting date will be chosen with a very limited window (in practice, it is restricted to a 6-week period starting in early December). Also, whether maize can be planted at any other time of year without negating most of the advantages of the velvetbean system is unclear. Another problem is the tough competition that the quickly reestablishing velvetbean gives to the growing maize crop in certain years, obliging farmers to prune the velvetbean. Finally, the year-to-year variability in the rate at which nutrients are made available via decomposition is difficult to predict (up to now at least), which may be seen as a constraint, especially if farmers make the achievement of high maize yields more and more a part of their agenda.

In sum, the analysis in this study suggests that the continuous annual rotation of velvetbean and maize can be sustained for at least 15 years at a reasonably high level of productivity (about 2–4 t ha^{-1} $year^{-1}$, with its current form of management), without any apparent decline in the natural resource base. Clearly, conserving or even improving the resource base does not in itself guarantee the global sustainability of a cropping system — an issue discussed at various points throughout the rest of this book. But the *abonera* system at least offers farmers the option of cultivating the same plot continuously if they wish to. If they decide to shift the land to another use, fields that have been in the velvetbean system for several years have none of the restrictive characteristics of degraded agricultural soils (low fertility, high weed or pest pressure, compaction, etc.). These fields are probably in an ideal condition to guarantee success with any other crop-, pasture-, or tree-based system.

The *abonera* system not only is an elegant way to provide and recycle nutrients or build up soil fertility, which has been the focus of this study, but also offers a host of other benefits, including weed control, reduced labour, and lower production risk. Clearly, the success of the *abonera* system varies among farmers and locations. For example, not all velvetbean fields accumulate quite enough biomass and nutrients to satisfy all maize nutritional requirements. Similarly, management decisions (from the choice of timing of slashing–planting to that of planting densities, as well as the timeliness of weed control) affect the production potential of the fields. This said, the fact remains that the performance of the *abonera* system would undoubtedly be much less satisfactory if not for its nutritional or soil-fertility benefits. These benefits, and their impacts on maize yields, also form the core of the economic analysis of the *abonera* system, to which we now turn.

Modern Agriculture: Ecological Impacts and the Possibilities for Truly Sustainable Farming

Until about four decades ago, crop yields in agricultural systems depended on internal resources, recycling of organic matter, built-in biological control mechanisms and rainfall patterns. Agricultural yields were modest, but stable. Production was safeguarded by growing more than one crop or variety in space and time in a field as insurance against pest outbreaks or severe weather. Inputs of nitrogen were gained by rotating major field crops with legumes. In turn rotations suppressed insects, weeds and diseases by effectively breaking the life cycles of these pests. A typical corn belt farmer grew corn rotated with several crops including soybeans, and small grain production was intrinsic to maintain livestock. Most of the labor was done by the family with occasional hired help and no specialized equipment or services were purchased from off-farm sources. In these type of farming systems the link between agriculture and ecology was quite strong and signs of environmental degradation were seldom evident.

But as agricultural modernization progressed, the ecology-farming linkage was often broken as ecological principles were ignored and/or overridden. In fact, several agricultural scientists have arrived at a general consensus that modern agriculture confronts an environmental crisis. A growing number of people have become concerned about the long-term sustainability of existing food production systems. Evidence has accumulated showing that whereas the present capital- and technology-intensive farming systems have been extremely productive and competitive, they also bring a variety of economic, environmental and social problems.

Evidence also shows that the very nature of the agricultural structure and prevailing policies have led to this environmental crisis by favoring large farm size, specialized production, crop monocultures and mechanization. Today as more and more farmers are integrated into international economies, imperatives to diversity disappear and monocultures are rewarded by economies of scale. In turn, lack of rotations and diversification take away key self-regulating mechanisms, turning monocultures into highly vulnerable agroecosystems dependent on high chemical inputs.

The Expansion of Monocultures

Today monocultures have increased dramatically worldwide, mainly through the geographical expansion of land devoted to single crops and year-to-year production of the same crop species on the same land. Available data indicate that the amount of crop diversity per unit of arable land has decreased and that croplands have shown a tendency toward concentration. There are political and economic forces influencing the trend to devote large areas to monoculture, and in fact such systems are rewarded by economies of scale and contribute significantly to the ability of national agricultures to serve international markets.

The technologies allowing the shift toward monoculture were mechanization, the improvement of crop varieties, and the development of agrochemicals to fertilize crops and control weeds and pests. Government commodity policies these past several decades encouraged the acceptance and utilization of these technologies. As a result, farms today are fewer, larger, more specialized and more capital intensive. At the regional level, increases in monoculture farming meant that the whole agricultural support infrastructure (i.e. research, extension, suppliers, storage, transport, markets, etc.) has become more specialized.

From an ecological perspective, the regional consequences of monoculture specialization are many-fold:

a. Most large-scale agricultural systems exhibit a poorly structured assemblage of farm components, with almost no linkages or complementary relationships between crop enterprises and among soils, crops and animals.

b. Cycles of nutrients, energy, water and wastes have become more open, rather than closed as in a natural ecosystem. Despite the substantial amount of crop residues and manure produced in farms, it is becoming increasingly difficult to recycle nutrients, even within agricultural systems. Animal wastes cannot economically be returned to the land in a nutrient-recycling process because production systems are geographically remote from other systems which would complete the cycle. In many areas, agricultural waste has become a liability rather than a resource. Recycling of nutrients from urban centers back to the fields is similarly difficult.

c. Part of the instability and susceptibility to pests of agroecosystems can be linked to the adoption of vast crop monocultures, which have concentrated resources for specialist crop herbivores and have increased the areas available for immigration of pests. This simplification has also reduced environmental opportunities for natural enemies. Consequently, pest outbreaks often occur when large numbers of immigrant pests, inhibited populations of beneficial insects, favorable weather and vulnerable crop stages happen simultaneously.

d. As specific crops are expanded beyond their "natural" ranges or favorable regions to areas of high pest potential, or with limited water, or low-fertility soils, intensified chemical controls are required to overcome such limiting factors. The assumption is that the human intervention and level of energy inputs that allow these expansions can be sustained indefinitely.

e. Commercial farmers witness a constant parade of new crop varieties as varietal replacement due to biotic stresses and market changes has accelerated to unprecedented levels. A cultivar with improved disease or insect resistance makes a debut, performs well for a few years (typically 5-9 years) and is then succeeded by another variety when yields begin to slip, productivity is threatened, or a more promising cultivar becomes available. A variety's trajectory is characterized by a take-off phase when it is adopted by farmers, a middle stage when the planted area stabilizes and finally a retraction of its acreage. Thus, stability in modern agriculture hinges on a continuous supply of new cultivars rather than a patchwork quilt of many different varieties planted on the same farm.

f. The need to subsidize monocultures requires increases in the use of pesticides and fertilizers, but the efficiency of use of applied inputs is decreasing and crop yields in most key crops are leveling off. In some places, yields are actually in decline. There are different opinions as to the underlying causes of this phenomenon. Some believe that yields are leveling off because the maximum yield potential of current varieties is being approached, and therefore genetic engineering must be applied to the task of redesigning crop. Agroecologists, on the other hand, believe that the leveling off is because of the steady erosion of the productive base of agriculture through unsustainable practices.

The First Wave of Environmental Problems

The specialization of production units has led to the image that agriculture is a modern miracle of food production. Evidence indicates, however, that excessive reliance on monoculture farming and agroindustrial inputs, such as capital-intensive technology, pesticides, and chemical fertilizers, has negatively impacted the environment and rural society. Most agriculturalists had assumed that the agroecosystem/natural ecosystem dichotomy need not lead to undesirable consequences, yet, unfortunately, a number of "ecological diseases" have been associated with the intensification of food production. They may be grouped into two categories: diseases of the ecotope, which include erosion, loss of soil fertility, depletion of nutrient reserves, salinization and alkalinization, pollution of water systems, loss of fertile croplands to urban development, and diseases of the biocoenosis, which include loss of crop, wild plant, and animal genetic resources, elimination of natural enemies, pest resurgence and genetic resistance to pesticides, chemical contamination, and destruction of natural control mechanisms. Under conditions of intensive management, treatment of such "diseases" requires an increase in the external costs to the extent that, in some agricultural systems, the amount of energy invested to produce a desired yield surpasses the energy harvested.

The loss of yields due to pests in many crops (reaching about 20-30% in most crops), despite the substantial increase in the use of pesticides (about 500 million kg of active ingredient worldwide) is a symptom of the environmental crisis affecting agriculture. It is well known that cultivated plants grown in genetically homogenous monocultures do not possess the necessary ecological defense mechanisms to tolerate the impact of outbreaking pest populations. Modern agriculturists have selected crops for high yields and high palatability, making them more susceptible to pests by sacrificing natural resistance for productivity. On the other hand, modern agricultural practices negatively affect pest natural enemies, which in turn do not find the necessary environmental resources and opportunities in monocultures to effectively and biologically suppress pests. Due to this lack of natural controls, an investment of about 40 billion dollars in pesticide control is incurred yearly by US farmers, which is estimated to save approximately $16 billion in US crops. However, the indirect costs of pesticide use to the environment and public health have to be balanced against these benefits. Based on the available data, the environmental (impacts on wildlife, pollinators, natural enemies, fisheries, water and development of resistance) and social costs (human poisonings and illnesses) of pesticide use reach about $8 billion each year. What is worrisome is that pesticide use is on the rise. Data from California shows that from 1941 to 1995 pesticide use increased from 161 to 212 million pounds of active ingredient. These increases were not due to increases in planted acreage, as statewide crop acreage remained constant during this period. Crops such as strawberries and grapes account for much of this increased use, which includes toxic pesticides, many of which are linked to cancers.

Fertilizers, on the other hand, have been praised as being highly associated with the temporary increase in food production observed in many countries. National average rates of nitrate applied to most arable lands fluctuate between 120-550 kg N/ha. But the bountiful harvests created at least in part through the use of chemical fertilizers, have associated, and often hidden, costs. A primary reason why chemical fertilizers pollute the environment is due to wasteful application and the fact that crops use them inefficiently. The fertilizer that is not recovered by the crop ends up in the environment, mostly in surface water or in ground water. Nitrate contamination of aquifers is widespread and in dangerously high levels in many rural regions of the world. In the US, it is estimated that more than 25% of the drinking water wells contain nitrate levels above the 45 parts per million safety standard. Such nitrate levels are hazardous to human health and studies have linked nitrate uptake to methaemoglobinemia in children and to gastric, bladder and oesophageal cancers in adults.

Fertilizer nutrients that enter surface waters (rivers, lakes, bays, etc.) can promote eutrophication, characterized initially by a population explosion of photosynthetic algae. Algal blooms turn the water bright green, prevent light from penetrating beneath surface layers, and therefore killing plants living on the bottom. Such dead vegetation serve as food for other aquatic microorganisms which soon deplete water of its oxygen, inhibiting the decomposition of organic residues, which accumulate on the bottom. Eventually, such nutrient enrichment of freshwater ecosystems leads to the destruction of all animal life in the water systems. In the US it is estimated that about 50-70% of all nutrients that reach surface waters is derived from fertilizers.

Chemical fertilizers can also become air pollutants, and have recently been implicated in the destruction of the ozone layer and in global warming. Their excessive use has also been linked to the acidification/salinization of soils and to a higher incidence of insect pests and diseases through mediation of negative nutritional changes in crop plants.

It is clear then that the first wave of environmental problems is deeply rooted in the prevalent socioeconomic system which promotes monocultures and the use of high input technologies and agricultural practices that lead to natural resource degradation. Such degradation is not only an ecological process, but also a social and political-economic process. This is why the problem of agricultural production cannot be regarded only as a technological one, but while agreeing that productivity issues represent part of the problem, attention to social, cultural and economic issues that account for the crisis is crucial. This is particularly true today where the economic and political domination of the rural development agenda by agribusiness has thrived at the expense of the interests of consumers, farmworkers, small family farms, wildlife, the environment, and rural communities.

The Second Wave of Environmental Problems

Despite that awareness of the impacts of modern technologies on the environment increased, as we traced pesticides in food chains and crop nutrients in streams and aquifiers, there are those that confronted to the challenges of the XXI century still argue for further intensification to meet the requirements of agricultural production. It is in this context that supporters of "status-quo agriculture" celebrate the emergence of biotechnology as the latest magic bullet that will revolutionize agriculture with products based on natures' own methods, making farming more environmentally friendly and more profitable for the farmer. Although clearly certain forms of non-transformational biotechnology hold promise for an improved agricult ıre, given its present orientation and control by multinational corporations, it holds more promise for environmental harm, for the further industrialization of agriculture and for the intrusion of private interests too far into public interest sector research.

What is ironic is the fact that the biorevolution is being brought forward by the same interests (Monsanto, Novartis, DuPont, etc.) that promoted the first wave of agrochemically-based agriculture, but this time, by equipping each crop with new "insecticidal genes", they are promising the world safer pesticides, reduction on chemically intensive farming and a more sustainable agriculture.

However, as long as transgenic crops follow closely the pesticide paradigm, such biotechnological products will do nothing but reinforce the pesticide treadmill in agroecosystems, thus legitimizing the concerns that many scientists have expressed regarding the possible environmental risks of genetically engineered organisms.

So far, field research as well as predictions based on ecological theory, indicate that among the major environmental risks associated with the release of genetically engineered crops can be summarized as follows:

- The trends set forth by corporations is to create broad international markets for a single product, thus creating the conditions for genetic uniformity in rural landscapes. History has repeatedly shown that a huge area planted to a single cultivar is very vulnerable to a new matching strain of a pathogen or pest;
- The spread of transgenic crops threatens crop genetic diversity by simplifying cropping systems and promoting genetic erosion;
- There is potential for the unintended transfer to plant relatives of the "transgenes" and the unpredictable ecological effects. The transfer of genes

from herbicide resistant crops (HRCs) to wild or semidomesticated relatives can lead to the creation of super weeds;

- Most probably insect pests will quickly develop resistance to crops with Bt toxin. Several Lepidoptera species have been reported to develop resistance to Bt toxin in both field and laboratory tests, suggesting that major resistance problems are likely to develop in Bt crops which through the continuous expression of the toxin create a strong selection pressure;
- Massive use of Bt toxin in crops can unleash potential negative interactions affecting ecological processes and non-target organisms. Evidence from studies conducted in Scotland suggest that aphids were capable of sequestering the toxin from Bt crops and transferring it to its coccinellid predators, in turn affecting reproduction and longevity of the beneficial beetles;
- Bt toxins can also be incorporated into the soil through leaf materials and litter, where they may persist for 2-3 months, resisting degradation by binding to soil clay particles while maintaining toxic activity, in turn negatively affecting invertebrates and nutrient cycling;
- A potential risk of transgenic plants expressing viral sequences derives from the possibility of new viral genotypes being generated by recombination between the genomic RNA of infecting viruses and RNA transcribed from the transgene;
- Another important environmental concern associated with the large scale cultivation of virus-resistant transgenic crops relates to the possible transfer of virus-derived transgenes into wild relatives through pollen flow.

Although there are many unanswered questions regarding the impact of the release of transgenic plants and micro-organisms into the environment, it is expected that biotechnology will exacerbate the problems of conventional agriculture and by promoting monocultures will also undermine ecological methods of farming such as rotations and polycultures. Because transgenic crops developed for pest control emphasize the use of a single control mechanism, which has proven to fail over and over again with insects, pathogens and weeds, transgenic crops are likely to increase the use of pesticides and to accelerate the evolution of "super weeds" and resistant insect pest strains. These possibilities are worrisome, especially when considering that during the period 1986-1997, approximately 25,000 transgenic crop field trials were conducted worldwide on more than 60 crops with 10 traits in 45 countries. By 1997 the global area devoted to transgenic crops reached 12.8 million hectares. Seventy-two percent of all transgenic crop field trials were conducted in the USA and Canada, although some were also conducted in descending order in Europe, Latin America and Asia. In most countries biosafety standards to monitor such releases are absent or are inadequate to predict ecological risks. In the industrialized countries from 1986-1992, 57% of all field trials to test transgenic crops involved herbicide tolerance pioneered by 27 corporations including the world's eight largest pesticide companies. As Roundup and other broad spectrum herbicides are increasingly deployed into croplands, the options for farmers for a diversified agriculture will be even more limited.

The Array of Alternatives to Conventional Agriculture.

Reduction and, especially, elimination of agrochemical require major changes in management to assure adequate plant nutrients and to control crop pests. As it was done a few decades ago, alternative sources of nutrients to maintain soil fertility include manures, sewage sludge and other organic wastes, and legumes in cropping sequences. Rotation benefits are due to biologically fixed nitrogen and from the interruption of weed, disease and insect cycles. A livestock enterprise may be integrated with grain cropping to provide animal manures and to utilize better the forages produced. Maximum benefits of pasture integration can be realized when livestock, crops, animals and other farm resources are assembled in mixed and rotational designs to optimize production efficiency, nutrient cycling and crop protection.

In orchards and vineyards, the use of cover crops improve soil fertility, soil structure and water penetration, prevent soil erosion, modify the microclimate and reduce weed competition. Entomological studies conducted in orchards with ground cover vegetation indicate that these systems exhibit lower incidence of insect pests than clean cultivated orchards. This is due to a higher abundance and efficiency of predators and parasitoids enhanced by the rich floral undergrowth.

Increasingly, researchers are showing that it is possible to provide a balanced environment, sustained yields, biologically mediated soil fertility and natural pest regulation through the design of diversified agroecosystems and the use of low-input technologies. Many alternative cropping systems have been tested, such as double cropping, strip cropping, cover cropping and intercropping, and more importantly concrete examples from real farmers show that such systems lead to optimal recycling of nutrients and organic matter turnover, closed energy flows, water and soil conservation and balanced pest-natural enemy populations. Such diversified farming exploit the complementarities that result from the various combinations of crops, trees and animals in spatial and temporal arrangements.

In essence, the optimal behavior of agroecosystems depends on the level of interactions between the various biotic and abiotic components. By assembling a functional biodiversity it is possible to initiate synergisms which subsidize agroecosystem processes by providing ecological services such as the activation of soil biology, the recycling of nutrients, the enhancement of beneficial arthropods and antagonists, and so on. Today there is a diverse selection of practices and technologies available, and which vary in effectiveness as well as in strategic value.

The Barriers for the Implementation of Alternatives

The agroecological approach seeks the diversification and revitalization of medium size and small farms and the reshaping of the entire agricultural policy and food system in ways that are economically viable to farmers and consumers. In fact, throughout the world there are hundreds of movements that are pursuing a change toward ecologically sensitive farming systems from a variety of perspectives. Some emphasize the production of organic products for lucrative markets, others land stewardship, while others the empowerment of peasant communities. In general, however, the goals are usually the same: to secure food self-sufficiency, to preserve the natural resource base, and to ensure social equity and economic viability.

What happens is that some well-intentioned groups suffer from "technological determinism", and emphasize as a key strategy only the development and dissemination of low-input or appropriate technologies as if these technologies in themselves have the capability of initiating beneficial social changes. The organic farming school that emphasizes input substitution (i.e. a toxic chemical substituted by a biological insecticide) but leaving the monoculture structure untouched, epitomizes those groups that have a relatively benign view of capitalist agriculture. Such perspective has unfortunately prevented many groups from understanding the structural roots of environmental degradation linked to monoculture farming.

This narrow acceptance of the present structure of agriculture as a given condition restricts the real possibility of implementing alternatives that challenge such a structure. Thus, options for a diversified agriculture are inhibited among other factors by the present trends in farm size and mechanization. Implementation of such mixed agriculture would only be possible as part of a broader program that includes, among other strategies, land reform and redesign of farm machinery adapted to polycultures. Merely introducing alternative agricultural designs will do little to change the underlying forces that led to monoculture production, farm size expansion, and mechanization in the first place.

Similarly, obstacles to changing cropping systems has been created by the government commodity programs in place these last several decades. In essence, these programs have rewarded those who maintained monocultures on their base feed grain acres by assuring these producers a particular price for their product. Those who failed to plant the allotted acreage of corn and other price-supported crops lost one deficit hectrage from their base. Consequently this created a competitive disadvantage for those who used a crop rotation. Such a disadvantage, of course, exacerbated economic hardship for many producers. Obviously many policy changes are necessary in order to create an economic scenario favorable to alternative cropping practices.

On the other hand, the large influence of multinational companies in promoting sales of agrochemicals cannot be ignored as a barrier to sustainable farming. Most

MNCs have taken advantage of existing policies that promote the enhanced participation of the private sector in technology development and delivery, positioning themselves in a powerful position to scale up promotion and marketing of pesticides. Realistically then the future of agriculture will be determined by power relations, and there is no reason why farmers and the public in general, if sufficiently empowered, could not influence the direction of agriculture along sustainability goals.

Clearly the nature of modern agricultural structure and contemporary policies have decidedly influenced the context of agricultural technology and production, which in turn has led to environmental problems of a first and second order. In fact, given the realities of the dominant economic milieu, policies discourage resource-conserving practices and in many cases such practices are not privately profitable for farmers. So the expectation that a set of policy changes could be implemented for a renaissance of diversified or small scale farms may be unrealistic, because it negates the existence of scale in agriculture and ignores the political power of agribusiness corporations and current trends set forth by globalization. A more radical transformation of agriculture is needed, one guided by the notion that ecological change in agriculture cannot be promoted without comparable changes in the social, political, cultural and economic arenas that also conform agriculture. In other words, change toward a more socially just, economically viable, and environmentally sound agriculture should be the result of social movements in the rural sector in alliance with urban organizations. This is especially relevant in the case of the new biorevolution, where concerted action is needed so that biotechnology companies feel the impact of environmental, farm labor, animal rights and consumers lobbies, pressuring them to re-orienting their work for the overall benefit of society and nature.

The Climate Change: Agriculture Conundrum

The risks associated with climate change lie in the interaction of several systems with many variables that must be collectively considered. Agriculture (including crop agriculture, animal husbandry, forestry and fisheries) can be defined as one of the systems, and climate the other. If these systems are treated independently, this would lead to an approach which is too fragmentary. The issue is more global. It is now held as likely that human activities can affect climate, one of the components of the environment. Climate in turn affects agriculture, the source of all food consumed by human beings and domestic animals. It must be further considered that not only climate may be changing, but that human societies and agriculture develop trends and constraints of their own which climate change impact studies must take into consideration.

An expert meeting held at FAO Headquarters in Rome from 7 to 10 December 1993 considered the direct effects of changing hydrological, pedological and plant physiological processes on agricultural production and concentrated on mechanisms.

Firstly, sustainability of agricultural and rural development. How will the links between environmental resources and demography be affected in the coming 50 years? Will it be possible, at the same time, to increase food production without irremediably losing environmental resources like soils or biodiversity?

Secondly, improved food security and nutrition, two members of a spiral which also includes rural poverty and demographic pressure. What are the prospects of breaking the vicious poverty circle in many developing countries under changing climate conditions?

The World Agricultural Context

In general, global food production has been growing faster than human population. There are, however, marked disparities between continents. For example, in Africa local production of cereals cannot keep pace with population increase, and production of root and tuber crops is growing faster than that of the nutritionally more valuable cereals.

In contrast, arable land growth lags behind population growth, which indicates some intensification of production. In Asia, the upper limit of available land has been reached in several countries, resulting in very high cropping intensities and a dominant role for irrigation. In Latin America, the increase in arable land is achieved only at a high ecological cost (especially deforestation) which may have direct relevance to climate change.

According to a recent FAO prospective study covering the years 1988/1990 to 2010 (*World Agriculture: Towards 2010 or AT-2010;* FAO, 1995a), 'the rate of growth of agricultural land will be further reduced during the next two decades. The pressures on *fresh* water *resources,* however, will be considerable, as will be those on the environment arising from the intensification of land use'.

A significant reduction in world population growth rates is foreseen from 1.8% (1980/1990) to 1.4% (2000/2010), roughly equivalent to an increase of the population doubling time from 40 to 50 years. The projected continuation of the high population growth in Africa can be related to slow economic development during the coming two decades, since in general a reduction of poverty precedes reduction of population growth.

AT-2010 also lists the following significant trends for the near future: (i) world production of cereals will continue to grow, but not in per capita terms; (ii) export cereals will undergo a modest growth in demand; (iii) the livestock sector in developing countries will continue to grow; (iv) root crops, tubers and plantains will retain their importance; (v) oil crops will undergo rapid growth in developing countries; and, most importantly, (vi) many developing countries will become net agricultural importers.

As indicated above for population increase, large differences exist in the rate of agricultural development among the developing continents, while differences among the developed countries tend to level out.

It is also likely that hunger and under-nutrition, which currently affects 800 million people (about 20% of the 4 000 million inhabitants of developing countries), will be reduced, but large pockets of malnutrition will persist. Pressure on environmental resources will continue to build up.

The Changing Agricultural Environment

The Climate 'Complex'

Climate constitutes a complex of inter-related variables. On average, through a set of regulatory mechanisms, a smooth change in one variable triggers smooth changes in most others. With the exception of possible qualitative and abrupt variations, which will be mentioned below, such inter-relations are independent of atmospheric carbon dioxide (CO_2). The latter and other greenhouse gases play a part largely through their effect on the radiation balance of the atmosphere.

There is only a weak link between such factors as cloudiness and wind. Temperature, evaporation and rain are strongly correlated, which illustrates the likely intensification of the hydrological cycle. Combined with the projected pressure on land and water use, competition for land and water will certainly become a key social and political issue.

Climate variability is likely to increase under global warming (Katz and Brown, 1992), both in absolute and in relative terms. This is linked with thresholds which affect the occurrence of many meteorological phenomena. For instance, tropical cyclones are 'fed' by water vapour evaporating from oceans at a temperature above 26 or 27°C. Therefore, higher average sea surface temperatures are bound to result in a higher frequency of tropical cyclones.

The rate of change itself is extremely important. For example, recent work (Lehman, 1993; Paillard and Labeyrie, 1994; Rahmsdorf, 1994; Holmes, 1995) on the saw-tooth temperature changes in the past as observed in the Arctic, raise concern that changes may occur abruptly, with average temperatures changing by 10 or 12°C in just a matter of decades. The mechanism of such changes is not clear as yet, but seems to involve the mechanical stability of ice sheets, and sudden changes in the hydrological cycle and the Atlantic conveyor current.

Such changes would, of course, be associated with dramatic changes in the distribution and quantities of ocean products, and cause havoc to established national fishery activities. They would also make adaptation to climate change, together with most agricultural planning, extremely difficult.

Agricultural Greenhouse Gases

In addition to water vapour, important greenhouse gases are carbon dioxide (CO_2), methane (CH_4), nitrous oxide (N_2O), tropospheric ozone (O_3) and chlorofluorocarbons (CFCs). The basic characteristics of the first three gases. The degree to which these greenhouse gases stem from agricultural sources is also given. The exact ratio between these land-use related emissions and those from natural ecosystems (swamps, tundra), fossil fuels (coal, oil, gas) and geological sources (volcanoes) are a bone of contention among industrialized, oil producing and developing countries.

Moral rights and duties in relation to the international flow of development and conservation funds are involved. Therefore, better estimates based on exact measurements of the net greenhouse gas emissions from agricultural practices in developing countries are urgently required to assess responsibilities properly. The required reductions of emissions to achieve stabilization of atmospheric concentrations of current levels are believed to be >60% for CO_2 15-20% for CH_4 and 70-80% for N_2O. This implies the need for up-to-date and complete information on land cover and land uses per country: not only the kind of crops grown but also their intensity, their rotation and the amount of inputs (energy, fertilizers).

The current trends of some of the agricultural sectors directly associated with greenhouse gas emission sources are listed. The loss of 'natural land', including tropical forests to agriculture, grazing, logging and urbanization, may continue, though at a slower pace (*AT-2010*). Many earlier estimates are now regarded with suspicion, as the borderline between crop agriculture, forest and cattle agriculture appears to be a fuzzy one.

Finally, both recent trends of fertilizer use and *AT-2010* projections indicate major increases in consumption. Even with appropriate measures to optimize fertilizer use, it is likely that N_2O losses from fertilizers will continue to increase.

Ecological and Indirect Climate Effects

In qualitative terms, many indirect effects of climate change on agriculture can be conjectured. Most of them are estimated to be negative and they catch most of the attention of the media. These effects include:

- the overall predictability of weather and climate would decrease, making the day-to-day and medium-term planning of farm operations more difficult;
- loss of biodiversity from some of the most fragile environments, such as tropical forests and mangroves;
- sea-level rise (40 cm in the coming 100 years) would submerge some valuable coastal agricultural land;

- the incidence of diseases and pests, especially alien ones, could increase;
- present (agro) ecological zones could shift in some cases over hundreds of kilometres horizontally, and hundreds of metres altitudinally, with the hazard that some plants, especially trees, and animal species cannot follow in time, and that farming systems cannot adjust themselves in time;
- higher temperatures would allow seasonally longer plant growth and crop growing in cool and mountainous areas, allowing in some cases increased cropping and production. In contrast, in already warm areas climate change can cause reduced productivity;
- the current imbalance of food production between cool and temperate regions and tropical and subtropical regions could worsen.

Plant Physiological Direct Effects

The greenhouse gases CH_4, N_2O and chlorofluorocarbons (CFCs) have no known direct effects on plant physiological processes. They only change global temperature and are therefore not discussed further. Instead, concentration should be on the effects of increased CO_2 tropospheric O_3, increased UV-B through depleted stratospheric ozone, increased temperatures and the associated intensification of the hydrological cycle.

Carbon Dioxide

The CO_2 Fertilization effect

CO_2 is an essential plant 'nutrient', in addition to light, suitable temperature, water and chemical elements such as N, P and K, and it is currently in short supply.

Higher concentrations of atmospheric CO_2 due to increased use of fossil fuels, deforestation

The Major Agricultural Crops and the Three Photosynthetic Pathways

- Plants are classified as C_3, C_4 or CAM according to the products formed in the initial phases of photosynthesis.
- C_3 species respond more to increased CO_2; C_4 species respond better than C_3 plants to higher temperature and their water-use efficiency increases more than for C_3 plants. There are some indications that enhancements can decline over time ('down-regulation').
- C_3 plants: cotton, rice, wheat, barley, soybeans, sunflower, potatoes, most leguminous and woody plants, most horticultural crops and many weeds
- C_4 plants: maize, sorghum, sugar cane, millets, halophytes (i.e., salt-tolerant plants) and many tall tropical grasses, pasture, forage and weed species

- CAM plants (Crassulacean Acid Metabolism, an optional C_3 or C_4 pathway of photosynthesis, depending on conditions): cassava, pineapple, opuntia, onions, castor

A total of 10 to 20% of the approximate doubling of crop productivity over the past 100 years could be due to this effect (Tans *et al.*, 1990) and forest growth or regrowth may have been stimulated as well. Further productivity increases may occur in the coming century, in the order of 30% or more where plant nutrients and moisture are adequate.

Higher CO_2 values would also mitigate the plant growth damage caused by pollutants such as NO_x and SO_2 because of smaller stomatal openings. Higher percentages of starch in grasses improves their feeding quality, implying less need for feed mixes when silaging.

The CO_2 Anti-transpirant Effect

With increased atmospheric CO_2 the consumptive use of water becomes more efficient because of reduced transpiration. This is induced by a contraction of plant stomata and/or a decrease in the number of stomata per unit leaf area. This restricts the escape of water vapour from the leaf more than it restricts photosynthesis.

With the same amount of available water, there could be more leaf area and biomass production by crops and natural vegetation. Plants could survive in areas hitherto too dry for their growth.

Ultraviolet Radiation

Increased ultraviolet radiation (UV-B, between 280 and 320 nanometres), due to depletion of the stratospheric ozone layer, mainly in the Antarctic region, may negatively affect terrestrial and aquatic photosynthesis and animal health. Over the last decade, a decrease of stratospheric ozone was observed at all latitudes (about 10% in winter, 0% during summer and intermediate values during spring and autumn). However, the 'Biological Action Factor' of UV-B can vary over several orders of magnitude with even slight changes in the amount and wavelength of UV-B.

The subject is treated in detail elsewhere in this book. In particular it can be noted that:

- there are damaging effects of increasing UV-B on crops, animals and plankton growth. It has been reported that UV-B affects the ability of plankton organisms to control their vertical movements and to adjust to light levels;
- reductions in yield of up to 10% have been measured at experimentally very high UV-B values, and would be particularly effective in plants where the

CO_2 fertilization effect is strongest. On the other hand, UV-B increase could increase the amount of plant internal compounds that act against pests.

Tropospheric Ozone

Tropospheric ozone originates about half from photochemical reactions involving nitrogen oxides (NO_x), methane or carbon monoxide, and half by downward movement of stratospheric ozone.

High ozone concentrations have toxic effects on both plant and animal life (German Bundestag, 1991; MacKenzie and El-Ashry, 1988; UNEP, 1993). It is likely that ozone, in conjunction with other photo-oxidants, is contributing towards the 'new type of forest damage' observed in Europe and the United States.

In the tropics, tropospheric ozone concentrations are generally lower than at northern mid-latitudes. However, this does not apply to periods when biomass burning releases precursor substances for the photochemical formation of ozone.

Temperature

In general, higher temperatures are associated with higher radiation and higher water use. It is relatively difficult to separate the physiological effects (at the level of plants and plant organs) of temperatures from the ecological ones (at the level of the field or of the region). There are both positive and negative impacts at the two levels, and only crop- and site-specific simulation can assess the global 'net' effect of temperature increases. It is generally agreed that:

- rising temperatures — now estimated to be 0.2°C per decade, or 1 °C by 2040 (Mitchell *et al.*, 1995) with smallest increases in the tropics (IPCC, 1992) — would diminish the yields of some crops, especially if night temperatures are increased (the temperature increase since the mid-1940s is mainly due to increasing night-time temperatures, while CO_2-induced warming would result in an almost equally large rise in minimum and maximum temperatures (Kukla and Karl, 1993);
- higher temperatures could have a positive effect on growth of plants of the CAM type. They would also strengthen the CO_2 fertilization effect and the CO_2 anti-transpirant effect of C_3 and C_4 plants unless plants get overheated;
- higher night temperature may increase dark respiration of plants, diminishing net biomass production;
- higher cold-season temperatures may lead to earlier ripening of annual crops, diminishing yield per crop, but would allow locally for the growth of more crops per year due to lengthening of the growing season. Winter kill of

pests is likely to be reduced at high latitudes, resulting in greater crop losses and higher need for pest control;

- higher temperatures will allow for more plant growth at high latitudes and altitudes.

Combined Effects and Some Uncertainties

The changes in CO_2 tropospheric ozone and increased UV-B do not necessarily occur simultaneously: CO_2 increase is worldwide, but with a strong seasonality in middle and higher latitudes; significant increase of UV-B is largely limited to subpolar regions (and mainly during the northern hemisphere winter months); high near-surface O_3 levels are restricted to the neighbourhood of major cities, airports, etc. (Seitz, 1994)

Box 1.2 illustrates some of the potential mechanisms which could account for either increased or decreased biomass under global change conditions. Note that increased biomass could even be associated with a decreased yield of grain (or sugar, oil, etc.) if one of the consequences of increased CO_2 will be a redistribution of biomass among plant organs (there are indications of relative increases of root growth).

The Hydrological Cycle and Soils

Even a slight increase in surface temperatures will affect evaporation, atmospheric moisture and precipitation. While it is generally agreed that rainfall will increase (by an estimated 10 to 15%), two aspects have to be elucidated: how will rainfall intensities be affected, and what are the details of spatial changes. This is still largely a matter of discussion among experts.

Based on palaeoclimatic analogies, certain authors predict more favourable rainfall conditions in the present-day Sahel (Petit-Maire, 1992). If the increase in precipitation should be associated with increased rainfall intensities, then the quality and quantity of soil and water resources would decline, for instance through increased runoff and erosion, increased land degradation processes, and a higher frequency of floods and possibly droughts. However:

- the extra precipitation on land, if indeed including present subhumid to semi-arid areas, will increase plant growth in these areas, leading to an improved protection of the land surface and increased rainfed agricultural production; in already humid areas the extra rainfall may, however, impair adequate crop drying and storage;
- the extra precipitation predicted to occur in some regions provides possibilities for off-site extra storage in rivers, lakes and artificial reservoirs (on-farm or at subcatchment level) for the benefit of improved rural water supply and expanded or more intensive irrigated agriculture and inland fisheries:

- the effects on water resources and water apportioning of international river and lake basins can be very substantial, with political overtones.

The greatest risks are often estimated to be associated with increased soil loss through erosion.

Soils, as a medium for plant growth, would be affected in several other ways:

- increased temperatures may lead to more decomposition of soil organic matter;
- increased plant growth due to the CO_2 fertilization effect may cause other plant nutrients such as N and P to become in short supply; however, CO_2 increase would stimulate mycorrhizal activity (making soil phosphorus more easily available), and also biological nitrogen fixation (whether or not symbiotic). Through increased root growth there would be extra weathering of the substratum, hence a fresh supply of potassium and micronutrients;
- the CO_2 fertilization effect would produce more litter of higher C/N ratio, hence more organic matter for incorporation into the soil as humus; litter with high C/N decomposes slowly and this can act as a negative feedback on nutrient availability;
- the 'CO_2 anti-transpirant' effect would stimulate plant growth in dryland areas, and more soil protection against erosion and lower topsoil temperatures, leading to an 'anti-desertification effect'.

Global climate change, if it occurs, will definitely affect agriculture. Most mechanisms, and two-way interactions between agriculture and climate, are known, even if not always well understood.

It is evident that the relationship between climate change and agriculture is still very much a matter of conjecture with many uncertainties; it remains largely a conundrum. Major uncertainties affect both the Global Circulation Models (GCMs) and the response of agriculture, as illustrated by differences among models, especially as regards effects at the national and subregional levels. In addition, many of the models do not take into consideration CO_2 fertilization and improved water-use efficiency, the effect of cloud cover (on both climate and photosynthesis), or the transient nature of climate change.

It is also worth remembering that enormous knowledge gaps still affect the carbon cycle (with a missing sink of about 2 Gt of carbon), the factors behind the recent near-stabilization of the atmospheric methane concentrations or the unexplained reduced rate of CO_2 increase in recent years, the effect of volcanic eruptions (such as the recent Pinatubo eruption), the effect of any increased cloudiness, etc.

Agroecology: Principles and Strategies for Designing Sustainable Farming Systems

The concept of sustainable agriculture is a relatively recent response to the decline in the quality of the natural resource base associated with modern agriculture (McIsaac and Edwards 1994). Today, the question of agricultural production has evolved from a purely technical one to a more complex one characterized by social, cultural, political and economic dimensions. The concept of sustainability although controversial and diffuse due to existing conflicting definitions and interpretations of its meaning, is useful because it captures a set of concerns about agriculture which is conceived as the result of the co-evolution of socioeconomic and natural systems (Reijntjes et al. 1992). A wider understanding of the agricultural context requires the study between agriculture, the global environment and social systems given that agricultural development results from the complex interaction of a multitude of factors. It is through this deeper understanding of the ecology of agricultural systems that doors will open to new management options more in tune with the objectives of a truly sustainable agriculture.

The sustainability concept has prompted much discussion and has promoted the need to propose major adjustments in conventional agriculture to make it more environmentally, socially and economically viable and compatible. Several possible solutions to the environmental problems created by capital and technology intensive farming systems have been proposed and research is currently in progress to evaluate alternative systems (Gliessman 1998). the main focus lies on the reduction or elimination of agrochemical inputs through changes in management to assure adequate plant nutrition and plant protection through organic nutrient sources and integrated pest management, respectively.

Although hundreds of more environmentally prone research projects and technological development attempts have taken place, and many lessons have been learned, the thrust is still highly technological, emphasizing the suppression of limiting factors or the symptoms that mask an ill producing agroecosystem. The prevalent philosophy is that pests, nutrient deficiencies or other factors are the cause of low productivity, as opposed to the view that pests or nutrients only become limiting if conditions in the agroecosystem are not in equilibrium (Carrol et al. 1990). For this reason, there still prevails a narrow view that specific causes affect productivity, and overcoming the limiting factor via new technologies, continues to be the main goal. This view has diverted agriculturists from realizing that limiting factors only represent symptoms of a more systemic disease inherent to unbalances within the agroecosystem and from an appreciation of the context and complexity of agroecological processes thus underestimating the root causes of agricultural limitations (Altieri et al. 1993).

On the other hand, the science of agroecology, which is defined as the application of ecological concepts and principles to the design and management of sustainable agroecosystems, provides a framework to assess the complexity of agroecosystems (Altieri 1995). The idea of agroecology is to go beyond the use of alternative practices and to develop agroecosystems with the minimal dependence on high agrochemical and energy inputs, emphasizing complex agricultural systems in which ecological interactions and synergisms between biological components provide the mechanisms for the systems to sponsor their own soil fertility, productivity and crop protection (Altieri and Rosset 1995).

Principles of Agroecology

In the search to reinstate more ecological rationale into agricultural production, scientists and developers have disregarded a key point in the development of a more self-sufficient and sustaining agriculture: a deep understanding of the nature of agroecosystems and the principles by which they function. Given this limitation, agroecology has emerged as the discipline that provides the basic ecological principles for how to study, design and manage agroecosystems that are both productive and natural resource conserving, and that are also culturally sensitive, socially just and economically viable (Altieri 1995).

Agroecology goes beyond a one-dimensional view of agroecosystems — their genetics, agronomy, edaphology, and so on,- to embrace an understanding of ecological and social levels of co-evolution, structure and function. Instead of focusing on one particular component of the agroecosystem, agroecology emphasizes the interrelatedness of all agroecosystem components and the complex dynamics of ecological processes (Vandermeer 1995).

Agroecosystems are communities of plants and animals interacting with their physical and chemical environments that have been modified by people to produce food, fibre, fuel and other products for human consumption and processing. Agroecology is the holitstic study of agroecosystems, including all environmental and human elements. It focuses on the form, dynamics and functions of their interrelationships and the processes in which they are involved. An area used for agricultural production, e.g. a field, is seen as a complex system in which ecological processes found under natural conditions also occur, e.g. nutrient cycling, predator/prey interactions, competition, symbiosis and successional changes. Implicit in agroecological research is the idea that, by understanding these ecological relationships and processes, agroecosystems can be manipulated to improve production and to produce more sustainably, with fewer negative environmental or social impacts and fewer external inputs (Altieri 1995).

The design of such systems is based on the application of the following ecological principles (Reinjntjes et al. 1992):

1. Enhance recycling of biomass and optimizing nutrient availability and balancing nutrient flow.
2. Securing favorable soil conditions for plant growth, particularly by managing organic matter and enhancing soil biotic activity.
3. Minimizing losses due to flows of solar radiation, air and water by way of microclimate management, water harvesting and soil management through increased soil cover.
4. Species and genetic diversification of the agroecosystem in time and space.
5. Enhance beneficial biological interactions and synergisms among agrobiodiversity components thus resulting in the promotion of key ecological processes and services.

These principles can be applied by way of various techniques and strategies. Each of these will have different effects on productivity, stability and resiliency within the farm system, depending on the local opportunities, resource constraints and, in most cases, on the market. The ultimate goal of agroecological design is to integrate components so that overall biological efficiency is improved, biodiversity is preserved, and the agroecosystem productivity and its self-sustaining capacity is maintained. The goal is to design a quilt of agroecosystems within a landscape unit, each mimicking the structure and function of natural ecosystems.

Biodiversification of Agroecosystems

From a management perspective, the agroecological objective is to provide a balanced environments, sustained yields, biologically mediated soil fertility and natural pest regulation through the design of diversified agroecosystems and the use of low-input technologies (Gleissman 1998). Agroecologists are now recognizing that intercropping, agroforestry and other diversification methods mimic natural ecological processes, and that the sustainability of complex agroecosystems lies in the ecological models they follow. By designing farming systems that mimic nature, optimal use can be made of sunlight, soil nutrients and rainfall (Pretty 1994).

Agroecological management must lead management to optimal recycling of nutrients and organic matter turnover, closed energy flows, water and soil conservation and balance pest-natural enemy populations. The strategy exploits the complementarities and synergisms that result from the various combinations of crops, tree and animals in spatial and temporal arrangements (Altieri 1994).

In essence, the optimal behavior of agroecosystems depends on the level of interactions between the various biotic and abiotic components. By assembling a functional biodiversity it is possible to initiate synergisms which subsidize agroecosystem processes by providing ecological services such as the activation of soil biology, the recycling of nutrients, the enhancement of beneficial arthropods and antagonists, and so on (Altieri and Nicholls 1999). Today there is a diverse selection of practices and technologies available, and which vary in effectiveness as well as in strategic value. Key practices are those of a preventative nature and which act by reinforcing the "immunity" of the agroecosystem through a series of mechanisms.

Various strategies to restore agricultural diversity in time and space include crop rotations, cover crops, intercropping, crop/livestock mixtures, and so on, which exhibit the following ecological features:

1. Crop Rotations. *Temporal diversity incorporated into cropping systems, providing crop nutrients and breaking the life cycles of several insect pests, diseases, and weed life cycles (Sumner 1982).*

2. *Polycultures*. Complex cropping systems in which tow or more crop species are planted within sufficient spatial proximity to result in competition or complementation, thus enhancing yields (Francis 1986, Vandermeer 1989).

3. *Agroforestry Systems*. An agricultural system where trees are grown together with annual crops and/or animals, resulting in enhanced complementary relations between components increasing multiple use of the agroecosystem (Nair 1982).

4. *Cover Crops*. The use of pure or mixed stands of legumes or other annual plant species under fruit trees for the purpose of improving soil fertility, enhancing biological control of pests, and modifying the orchard microclimate (Finch and Sharp 1976).

5. Animal integration in agroecosystems aids in achieving high biomass output and optimal recycling (Pearson and Ison 1987).

All of the above diversified forms of agroecosystems share in common the following features (Altieri and Rosset 1995):

a. Maintain vegetative cover as an effective soil and water conserving measure, met through the use of no-till practices, mulch farming, and use of cover crops and other appropriate methods.

b. Provide a regular supply of organic matter through the addition of organic matter (manure, compost, and promotion of soil biotic activity).

c. Enhance nutrient recycling mechanisms through the use of livestock systems based on legumes, etc.

d. Promote pest regulation through enhanced activity of biological control agents achieved by introducing and/or conserving natural enemies and antagonists.

Research on diversified cropping systems underscores the great importance of diversity in an agricultural setting (Francis 1986, Vandermeer 1989, Altieri 1995). Diversity is of value in agroecosystems for a variety of reasons (Altieri 1994, Gliessman 1998):

- As diversity increases, so do opportunities for coexistence and beneficial interactions between species that can enhance agroecosystem sustainability.
- Greater diversity often allows better resource-use efficiency in an agroecosystem. There is better system-level adaptation to habitat heterogeneity, leading to complementarity in crop species needs, diversification of niches, overlap of species niches, and partitioning of resources.
- Ecosystems in which plant species are intermingled possess an associated resistance to herbivores as in diverse systems there is a greater abundance and diversity of natural enemies of pest insects keeping in check the populations of individual herbivore species.
- A diverse crop assemblage can create a diversity of microclimates within the cropping system that can be occupied by a range of noncrop organisms — including beneficial predators, parasites, pollinators, soil fauna and antagonists — that are of importance for the entire system.
- Diversity in the agricultural landscape can contribute to the conservation of biodiversity in surrounding natural ecosystems.
- Diversity in the soil performs a variety of ecological services such as nutrient recycling and detoxification of noxious chemicals and regulation of plant growth.
- Diversity reduces risk for farmers, especially in marginal areas with more unpredictable environmental conditions. If one crop does not do well, income from others can compensate.

Agroecology and the Design of Sustainable Agroecosystems

Most people involved in the promotion of sustainable agriculture aim at creating a form of agriculture that maintains productivity in the long term by (Pretty 1994, Vandermeer 1995):

- optimizing the use of locally available resources by combining the different components of the farm system, i.e. plants, animals, soil, water, climate and people, so that they complement each other and have the greatest possible synergetic effects;
- reducing the use of off-farm, external and non-renewable inputs with the greatest potential to damage the environment or harm the health of farmers

and consumers, and a more targeted use of the remaining inputs used with a view to minimizing variable costs;

- relying mainly on resources within the agroecosystem by replacing external inputs with nutrient cycling, better conservation, and an expanded use of local resources;
- improving the match between cropping patterns and the productive potential and environmental constraints of climate and landscape to ensure long-term sustainability of current production levels;
- working to value and conserve biological diversity, both in the wild and in domesticated landscapes, and making optimal use of the biological and genetic potential of plant and animal species; and
- taking full advantage of local knowledge and practices, including innovative approaches not yet fully understood by scientists although widely adopted by farmers.

Agroecology provides the knowledge and methodology necessary for developing an agriculture that is on the on e hand environmentally sound and on the other hand highly productive, socially equitable and economically viable. Through the application of agroecological principles, the basic challenge for sustainable agriculture to make better use of internal resources can be easily achieved by minimizing the external inputs used, and preferably by regenerating internal resources more effectively through diversification strategies that enhance synergisms among key components of the agroecosystem.

The ultimate goal of agroecological design is to integrate components so that overall biological efficiency is improved, biodiversity is preserved, and the agroecosystem productivity and its self-regulating capacity is maintained. The goal is to design an agroecosystem that mimics the structure and function of local natural ecosystems; that is, a system with high species diversity and a biologically active soil, one that promotes natural pest control, nutrient recycling and high soil cover to prevent resource losses.

Agroecology provides guidelines to develop diversified agroecosystems that take advantage of the effects of the integration of plant and animal biodiversity such integration enhances complex interactions and synergisms and optimizes ecosystem functions and processes, such as biotic regulation of harmful organisms, nutrient recycling, and biomass production and accumulation, thus allowing agroecosystems to sponsor their own functioning. The end result of agroecological design is improved economic and ecological sustainability of the agroecosystem, with the proposed management systems specifically in tune with the local resource base and operational framework of existing environmental and socioeconomic conditions. In an agroecological strategy, management components are directed to highlight the conservation and enhancement of local agricultural resources (germplasm, soil, beneficial fauna, plant biodiversity, etc.) by emphasizing a development methodology that encourages farmer participation, use of traditional knowledge, and adaptation of farm enterprises that fit local needs and socioeconomic and biophysical conditions.

Ecological Processes to Optimize in Agroecosystems

- Strengthen the immune system (proper functioning of natural pest control)
- Decrease toxicity through elimination of agrochemicals
- Optimize metabolic function (organic matter decomposition and nutrient cycling)
- Balance regulatory systems (nutrient cycles, water balance, energy flow, population regulation, etc.)
- Enhance conservation and regeneration of soil-water resources and biodiversity
- Increase and sustain long-term productivity

Mechanisms to Improve Agroecosystem Immunity

- Increase of plant species and genetic diversity in time and space.
- Enhancement of functional biodiversity (natural enemies, antagonists etc.)
- Enhancement of soil organic matter and biological activity
- Increase of soil cover and crop competitive ability
- Elimination of toxic inputs and residues

The CO_2 Fertilization Effect: Higher Carbohydrate Production and Retention as Biomass and Seed Yield

The rise in atmospheric carbon dioxide (CO_2) concentration from about 280 m mol/mol before the industrial revolution to about 360 m mol/mol currently is well documented (e.g., Baker and Enoch, 1983; Keeling *et al.*, 1995). The consensus of many studies of the effects of elevated CO_2 on plants is that the CO_2 fertilization effect is real (see Kimball, 1983; Acock and Allen, 1985; Cure and Acock, 1986; Allen, 1990; Rozema *et al.*, 1993; Allen, 1994; Allen and Amthor, 1995). However, the CO_2 fertilization effect may not be manifested under conditions where some other growth factor is severely limiting, such as low temperature (Long, 1991). Also, plants grown in some conditions, where limitations of rooting volume (Arp, 1991), light, or other factors restrict growth, have not shown a sustained response to elevated CO_2 (Kramer, 1981).

The main objectives of this study are to assess the direct effects of rising atmospheric CO_2 and indirect effects of potential climate changes on crop growth and yield. The approach will be to (1) provide a general overview of CO_2 effects on plant growth processes; (2) analyse some specific experimental data on crop plant responses to elevated CO_2 and climate change factors; (3) summarize recent reviews of plant responses to elevated CO_2; and (4) discuss some crop modelling assessments of rising CO_2 and climate change factors on agricultural productivity based on predictions of global climate change models. Also, some adaptations for improving crop productivity in a higher CO_2 world will be suggested.

Overview of CO_2 Effects on Plant Growth Processes

Most of the following discussion of CO_2 effects on plants applies to species with the C_3 photosynthetic pathway and not necessarily to species with the C_4 pathway. Other aerial, non-biotic environmental factors that affect plant growth and development are light and temperature. Plant photosynthetic rates generally increase linearly with light across relatively low ranges of light intensity, and then the rates decelerate until they reach an asymptotic maximum. Because of crowding and shading of many leaves, most crop canopies do not reach light saturation at full sunlight; that is, they would be able to respond to light levels well beyond full solar irradiance. Likewise, crop photosynthetic rates respond to increasing levels of CO_2 but then level off at higher concentrations (around 700 m mol/mol or greater, depending upon species and other factors). However, leaf photosynthesis usually increases with temperature up to some maximum value, and then declines. Furthermore, temperature affects not only photosynthesis, but also respiration, growth, development phases and reproductive processes.

Elevated CO_2 may have some effects on crop phenology, although stages of development are governed primarily by temperature, time and photoperiod. If dates of planting were to be changed because of the greenhouse effect, then phenological timing of plants could be affected. For example, higher temperatures could decrease yields by decreasing the duration of the grain-filling period or changes in photoperiod could shorten or lengthen the vegetative stage.

The CO_2 fertilization effect begins with enhanced photosynthetic CO_2 fixation. Non-structural carbohydrates tend to accumulate in leaves and other plant organs as starch, soluble carbohydrates or polyfructosans, depending on species. In some cases, there may be feedback inhibition of photosynthesis associated with accumulation of non-structural carbohydrates. Increased carbohydrate accumulation, especially in leaves, may be evidence that crop plants grown under CO_2 enrichment may not be fully adapted to take complete advantage of elevated CO_2. This may be because the CO_2-enriched plants do not have an adequate sink (inadequate growth capacity), or lack capacity to load phloem and translocate soluble carbohydrates. Improvement of photoassimilate utilization should be one goal of designing cultivars for the future (Hall and Allen, 1993).

In the process of growth, photoassimilates are allocated to the vegetative shoots, root system or reproductive organs. In some cases, more photoassimilate of CO_2-enriched plants is partitioned to the root system than to the shoots. Above ground, more photoassimilate usually goes into stems and supporting structures than into leaves. This phenomenon may not be an inherent response to elevated CO_2, but may be a by-product of the larger size of plants often found in CO_2-enriched atmospheres, especially by species that produce branch stems along the aerial mainstems (Allen *et al.,* 1991).

Reproductive biomass growth as well as vegetative biomass growth are usually increased by elevated CO_2. However, the harvest index, or the ratio of seed yield to above-ground biomass yield, is typically lower under elevated CO_2 conditions (Allen, 1991; Baker *et al.,* 1989), which may also be evidence of the lack of capacity to utilize completely the more abundant photoassimilate.

In many cases, both the amount and the carboxylation activity of ribulose 1,5-bisphosphate carboxylase-oxygenase enzyme (rubisco) is decreased in leaves of plants grown under elevated CO_2 This acclimation phenomenon may produce 'downregulation of photosynthesis'; however, this is not universally the case. For example, there is little evidence of this downregulation response in soybean (*Glycine max* L. Merr.), a C_3 legume. In fact, photosynthetic capacity per unit leaf area of soybean is increased under CO_2 enrichment. Leaves often develop an additional layer of mesophyll cells. Also, more structural carbohydrate may be produced in leaves, as well as stems, of CO_2-enriched plants.

We can obtain clues about reasons that some plants downregulate and others do not in response to elevated CO_2 by focusing on the capabilities of soybean. This plant has (a) symbiotic N2 fixation; (b) the capacity to form additional layers of palisade cells in the leaf tissue; (c) the capacity to shunt much of the photoassimilate into relatively inert starch rather than soluble sugars during photosynthesis; (d) a relatively strong leaf and stem sink during vegetative development; and (e) a strong seed-fill sink during reproductive development. Plants which lack these capacities, either inherently or because of growth in limiting environments, are more likely to demonstrate some degree of downregulation of photosynthesis (Allen, 1994).

The carbon:nitrogen ratio of leaves of plants is usually increased under CO_2 enrichment. Plants may acclimate to elevated CO_2 by requiring less rubisco and photosynthetic apparatus, which would lead to lower nitrogen contents. The overall change in C:N ratios is governed both by increases in structural and non-structural carbohydrates, and by decreases in protein content. However, seed nitrogen content is little affected (Allen *et al.,* 1988).

Specific respiration rates may be reduced by both short-term exposure to elevated

CO_2 and long-term growth at elevated CO_2 (Amthor, 1995). However, the long-term effect may be similar when respiration rates are reported on a per unit nitrogen basis.

In climate change scenarios, temperatures are predicted to increase following the rise of CO_2 and other greenhouse-effect gases. Carbon dioxide x temperature interactions have been observed for vegetative growth (i.e., the CO_2 fertilization effect is greater at warmer temperatures than at cooler temperatures). Temperature increases in a higher CO_2 world could increase overall biomass productivity for vegetative crops (pastures and forages) both by extending the length of the growing season in temperate regions, and by the interaction of CO_2 x temperature in stimulation of vegetative growth. However, CO_2 x temperature interactions appear to be very small or negligible for reproductive processes (seed set and seed yield) although there may be more initial flowers formed by greater amounts of branching or tillering that is stimulated by CO_2 enrichment (Baker and Allen, 1993a).

Precipitation changes may occur along with other climatic change effects. In general, predictions from crop models show that increased CO_2 should increase productivity of C_3 plants, but the associated predictions of temperature rise will be detrimental. Not surprisingly, changes in precipitation patterns (decreases of rainfall during growth period) could be more detrimental for crop production than changes in temperature.

Under elevated CO_2 stomatal conductance in most species will decrease which may result in less transpiration per unit leaf area. However, leaf area index of some crops may also increase. The typical 40% reduction in stomatal conductance induced by a doubling of CO_2 has generally resulted in only a 10% (or less) reduction in crop canopy water use in chamber or field experimental conditions. Actual changes in crop evapotranspiration will be governed by the crop energy balance, as mitigated by stomatal conductance, leaf area index, crop structure and any changing meteorological factors.

Water-use efficiency (WUE) (ratio of CO_2 uptake to evapotranspiration) will increase under higher CO_2 conditions. This increase is caused more by increased photosynthesis than it is by a reduction of water loss through partially closed stomata. Thus, more biomass can be produced per unit of water used, although a crop would still require almost as much water from sowing to final harvest. If temperatures rise, however, the increased WUE caused by the CO_2 fertilization effect could be diminished or negated, unless planting dates can be changed to more favourable seasons.

Several assessments of impacts of climate change on crop productivity have been published. Progress has been made on integrating the impacts on individual countries and on economic and social interactions. For the most part, these assessments project more favourable climates for agriculture in northern latitudes and less favourable climates in the tropical and subtropical zones. However, the crop modelling predictions are dependent on the scenarios of outputs of General Circulation Models (GCMs) applied to the greenhouse effect. Thus, the dependency chain of assessments follows: Climate Change Scenarios → Crop Model Prediction and Agricultural Production Systems (with and without available mitigation and adaptation response strategies) → National Scenarios of Economics and Well-being of Farmers, Agricultural Commerce, and Consumers → Country-by-Country and Global Interaction Scenarios of World Trade (food and all other commodities). Population Dynamics and Economic Well-being, and Impacts on Social Systems.

As the world continues to consume fossil fuels, CO_2 concentrations will continue to rise. Other greenhouse-effect gases, such as methane, nitrous oxides, chlorofluorocarbons and chlorofluorocarbon substitutes, and perhaps tropospheric ozone, will likely rise also. The CO_2 fertilization effect on plants will increase and climate changes may occur because of the combined increase of all greenhouse-effect gases. Global agriculture could adapt to gradual regional climate changes, but sudden changes would be more serious. Adaptation and/or mitigation actions could include the following:

1. Selection of plants that can better utilize carbohydrates which are produced when plants are grown at elevated CO_2.
2. Selection of plants that produce less structural matter and more reproductive capacity under CO_2 enrichment. (This applies for seed crop plants, not necessarily vegetative biomass plants.)
3. Search for germplasms that are adapted to higher day and night temperatures, and incorporate those traits into desirable crop production cultivars to improve flowering and seed set.
4. Change planting dates and other crop management procedures to optimize yields under new climatic conditions, and select for cultivars that are adapted to these changed agricultural practices.
5. Shift to species that have more stable production under high temperatures or drought.
6. Determine whether more favourable N:C ratios can be attained in forage cultivars adapted to elevated CO_2.
7. Where needed, and where possible, develop irrigation systems for crops.

Specific Responses of Crops to Elevated CO_2

This section will focus on responses of two C_3 crops to elevated CO_2: soybean, a symbiotic nitrogen-fixing legume representative of the pulses; and rice (*Oryza sativa* L.), a globally important cereal food crop. Findings from seven crop cycles of both soybean and rice grown in sunlit chambers at Gainesville, Florida, USA, will be emphasized (Baker and Allen, 1993a,b). The chambers feature both large rooting volumes of real soil and light from the sun. Some studies included responses to sub-ambient as well as superambient CO_2 concentrations (e.g., Allen *et al.*, 1991; Baker and Allen, 1993a,b, 1994). (The rice experiments have been designated chronologically as RICE I, II, III, IV, V, VI and VII.)

Soyabean

Photosynthetic Rates

Soybean photosynthetic rates of both leaves and the whole canopy have *always* been increased by elevated CO_2 Jones *et al.* (1984) reported midday maximum canopy photosynthetic rates of 60 and 90 m mol $CO_2/m^2/s$ at 70 days after planting (DAP) for soybean grown at CO_2 concentrations of 330 and 800 m mol/mol, respectively. The leaf area index (LAI) was 6.9 and 9.0, respectively. Midday maximum canopy photosynthetic rates in other studies were 40 and 75 m mol/m^2/s for soybean grown at 320 and 640 m mol/mol CO_2 (Jones *et al.*, 1985c) and 40 and 80 m mol/m^2/s for soybean grown at 330 and 800 m mol/m^2/s (Jones *et al.*, 1985a,b).

Valle *et al.* (1985) found that midday maximum photosynthetic CO_2 uptake rates of soybean leaves ranged from 30 to 50 m mol/m^2/s and 15 to 25 m mol/m^2/s on plants grown at 660 and 330 m mol/mol CO_2 respectively. Allen *et al.* (1990) reported that, at all light levels, leaf photosynthetic rates increased linearly with CO_2 concentration across the range of 330 to 800 m mol/mol.

Valle *et al.* (1985) used a Michaelis-Menten type of rectangular hyperbola to summarize photosynthetic responses of soybean leaves vs. CO_2 concentration. The plants had been grown at 330 and 660 m mol/mol of CO_2 and then exposed to a wide range of CO_2 for a short period.

$$Y=(Y_{max} \times [C])/([C]+K_m)+Y_i \quad (4.1)$$

where Y is photosynthetic rate in m mol $CO_2/m^2/s$; $[C_3]$ is CO_2 concentration in m mol/mol; Y_i is the y-axis intercept at zero $[C_3]$, the apparent respiration rate, in m mol $CO_2/m^2/s$; Y_{max} is the response limit of ($Y — Y_i$) at very high [C], the asymptotic photosynthetic rate, in m mol $CO_2/m^2/s$; K_m is the value of [C] where $(Y-Y_i)=Y_{max}/2$, the apparent Michaelis-Menten constant, in (m mol/mol; and G_c is the calculated [C] intercept at zero Y, the CO_2 compensation point, m mol/mol (not shown in this equation). The average parameters for responses at 330 and 660 m mol/mol are given. There was no obvious downregulation of soybean leaf photosynthesis in response to elevated CO_2; in fact, photosynthetic capacity was increased. Leaf quantum yield increased from 0.05 to 0.09 in the soybean leaves exposed to CO_2 of 330 and 660 m mol/mol, respectively (Valle *et al.*, 1985).

Campbell *et al.* (1988) showed that soybean leaf photosynthetic rates were higher for plants grown at 660 than at 330 m mol/mol CO_2 when measured at common intercellular CO_2 concentrations. Furthermore, Campbell *et al.* (1988) measured rubisco activity and amount in leaves of soybean grown in CO_2 concentrations of 160, 220, 280, 330, 660 and 990 m mol/mol. They found that rubisco activity was almost constant at 1.0 m mol CO_2/min/mg soluble protein across this CO_2 treatment range. Leaf soluble protein was nearly constant at about 2.4 g/m^2 with 55% being rubisco protein. Specific leaf weight increased across the 160 to 990 m mol/mol CO_2 concentration range, so that the rubisco activity on a leaf dry weight basis decreased.

Campbell *et al.* (1990) showed that photosynthetic capacity of soybean canopies grown at 330 and 660 m mol/mol CO_2 were similar for short exposures at concentrations below 500 m mol/mol, but above this concentration, canopies grown at 660 m mol/mol had a slightly higher photosynthetic capacity. Thus, soybean did not lose photosynthetic capacity as did some other plant species. In other studies, however, both photosynthetic rate and rubisco activity of soybean declined during long-term CO_2 enrichment (Thorne and Koller, 1974; Delucia *et al.*, 1985).

Respiration

Whole-canopy pre-dawn respiration rates at 40 to 60 DAP were 2-3 and 4-5 m mol/m^2/s for soybean grown at CO_2 levels of 320 and 640 m mol/m^2, respectively, and a constant day/night temperature of 25°C (Jones *et al.*, 1985c). After cross-switching two of the treatment chambers at DAP 52, the subsequent canopy photosynthetic rates and respiration rates quickly adjusted to their new CO_2 exposure conditions.

Thus, pre-dawn respiration rates were closely connected to the previous CO_2 fixation rates.

Partitioning

Growth of plants under elevated CO_2 results in changes in partitioning of photoassimilates to various plant organs over time. In soybean, elevated CO_2 generally promoted greater carbon (dry matter) partitioning to the supporting structure (stems, petioles and roots) than to the leaf laminae during vegetative stages of growth (Allen *et al.*, 1991). During reproductive stages, there tended to be lower relative partitioning to reproductive growth (pods) by plants under elevated CO_2.

Growth Rates

During the linear phase of vegetative growth after full ground cover is reached, the growth rates of plants exposed to a range of CO_2 concentrations varied from 5.0 to 20.7 g/m^2/d for exposures from 160 to 990 m mol/mol (Allen *et al.*, 1991). The total final dry weight ranged from 12.88 to 39.12 g/plant, and final seed weight ranged from 5.77 to 17.85 g/plant for CO_2 treatments ranging from 160 to 660 m mol/mol.

Carbohydrates

Soybean accumulates non-structural carbohydrates, particularly starch, under CO_2 enrichment. Allen *et al.* (1988) grew soybean under CO_2 treatments of 330, 450, 600 and 800 m mol/mol. Sucrose, reducing sugars, and total soluble sugars of leaves remained somewhat constant throughout the day, but starch increased steadily at a rate of about 6 g/kg dry matter/hour. Average total soluble sugars increased from 24 to 36 g/kg dry weight and starch increased from 85 to 204 g/kg dry weight across the range of 330 to 800 m mol/mol. Elevated non-structural carbohydrates in CO_2-enriched soybean plants were confirmed by Baker *et al.* (1989) and Allen *et al.* (1995). The concentrations also varied across the life cycle of the plants.

Nitrogen

Nitrogen content decreased from 50 to 37 g/kg dry weight over the range of 330 to 800 m mol/mol. When the nitrogen content was adjusted to remove the effect of total non-structural carbohydrate, the relative changes were smaller, 55 to 48 g N/kg dry matter, across the 330 to 800 m mol/mol range (Allen *et al.,* 1988).

Yield

Soybean seed yield was always increased by elevated CO_2 Allen *et al.* (1987) summarized the photosynthetic, biomass and seed yield responses of several experiments with the equation (4.1) rectangular hyperbola model using data normalized to responses obtained at 330 m mol/mol. The values of K_m, Y_{max} and Y_i parameters for relative photosynthetic rates were 279 m mol/mol, 3.08 and -0.68, respectively, for relative biomass yield were 182 m mol/mol, 3.02 and -0.91, respectively, and for relative seed yield were 141 m mol/mol, 2.55 and -0.76, respectively. This model was used to project yields across several ranges of atmospheric CO_2 concentration increases. For a doubling of CO_2 this model predicted a 32.2% increase in soybean grain yield and a 42.7% increase in biomass. The ratio of these two numbers, 1.322/1.427 = 0.926, gives the fraction of the harvest index expected under doubled CO_2 in comparison with ambient CO_2.

Carbon Dioxide and Temperature

The CO_2 fertilization effect appears to be enhanced under elevated temperatures, at least up to a point. Idso *et al.* (1987) and Kimball *et al.* (1993) showed that the growth modification factor (or biomass growth modification ratio) due to a 300 m mol/mol enrichment was 0.08 per °C (average daily temperature) across the range of 12 to 34°C. However, for soybean, Allen (1991) calculated a biomass growth modification ratio response to temperature of -0.031 per °C for seed biomass yield and -0.026 per °C for total biomass accumulation. Elevated temperatures tended to shorten the grain-filling period of this crop.

Soybean seed yield tended to decrease slightly with temperature over the day/night range of 26/19 to 36/29°C. The number of seed per plant increased slightly with increase of both CO_2 and temperature. Mass per seed decreased sharply with increasing temperature. Although CO_2 enrichment resulted in increased seed yield and above-ground biomass, harvest index was decreased with both CO_2 and temperature (Baker *et al.*, 1989). The growth modification factor to increase with temperature for either seed yield or biomass accumulation.

Subsequent experiments with soybean substantiate these data although it was found that seed yields dropped sharply for day/night temperatures of 40/30°C and above, and biomass yields were maintained up to 44/34°C before rapidly failing (Deyun Pan *et al.*, unpublished).

The sensitivity of the growth modification factor data of Idso *et al.* (1987) and Kimball *et al.* (1993) may have been affected by other factors at Phoenix, Arizona. Apparently, the growth vs. temperature data were obtained throughout the year, and the findings may have been impacted by factors such as total solar radiation, photoperiod, stages of development, or other conditions caused by the changing seasons. Of course, these other environmental factors are all part of the complex of plant responses to climatic conditions. However, a better test of temperature effects alone would be CO_2 enrichment throughout the season or life cycle of plants under natural conditions, but with consistent temperature differences (cooler and warmer).

Evapotranspiration and Water-use Efficiency

The direct effect of increasing temperatures across the range of 28 to 35°C appears to increase transpiration rate about 4 to 5% per °C, based on both experimental and modelling studies (Allen, 1991). This is in close agreement with the rise in saturation vapour pressure of about 6% per °C. Allen *et al.* (1985) showed that the increase in WUE of soybean contributed by increased photosynthesis under elevated CO_2 was much greater than the increase in WUE contributed by decreased transpiration. The fractional increase in WUE attributable to increased photosynthesis and decreased transpiration were about 0.8 and 0.2, respectively (based on data of Jones *et al.*, 1985b).

Rice

Leaf and Canopy Photosynthesis

Leaf photosynthetic rates, in high light, of cv. IR72 rice, determined in 1992 (RICE VI experiment) for plants grown at three day/night temperature regimes, 32/23, 35/26 and 38/29°C (but measured at ambient temperature), averaged 18.8 and 30.4 m mol $CO_2/m^2/s$ for 330 and 660 m mol/mol CO_2 treatments, respectively (Allen *et al.*, 1995).

This increase caused by elevated CO_2 is about 60%. Leaf A/C_i curves obtained from both CO_2 treatments at the growth temperature of 35/26°C. There were essentially no differences in A between the two CO_2 treatments at each specific C. level. Similar pairs of curves (not shown) were obtained for the other temperature treatments. Each pair of leaf A/C_i curves were similar within each temperature regime. Thus, the leaf A/C_i curves gave no indication of a change in photo-synthetic capacity. Extrapolations of the response curves to zero values of A gave an intercept (leaf CO_2 compensation point) of about 60 m mol/mol CO_2.

Following canopy closure, the rice canopy net photosynthetic rate (Pn) vs. photosynthetic photon flux density (PPFD) responses were linear and did not approach light saturation in any of the experiments, probably because of the erect leaf orientation of cv. IR30 and high plant populations. Canopy Pn vs. PPFD at 60 DAP for the six CO_2 treatments of the RICE II experiment (Baker and Allen, 1993b; Allen *et al.*, 1995) gave values of about 34, 50, 60, 80, 85 and 90 m mol CO_2/m^2/s at high light (1600 m mol photons/m^2/s) for treatments of 160, 250, 330, 500, 660 and 900 m mol/mol.

Linear canopy Pn responses to PPFD on day 60 of the RICE IV experiment were similar among all temperature treatments (25/18, 28/21, 31/24, 34/27 and 37/30°C) that were exposed to 660 m mol/mol *CO_2* responses of the 330 m mol/mol treatment were about 25% less. Other studies of canopy Pn have shown only small differences across wide ranges of temperature; e.g., cotton (Baker *et al.*, 1972) and soybean (Jones *et al.*, 1985a). There may be two reasons for the lack of a clear canopy photosynthetic response to air temperature across the 25 to 37°C range. Evaporative cooling may lower foliage temperature below air temperature increasingly with increasing air temperature and increasing vapour pressure deficit (Allen, 1990; Pickering *et al.*, 1995). The summed response of the photosynthetic rates of leaves to temperature at all exposures of light may broaden the temperature response for the whole canopy photo-synthetic rates (Pickering *et al.*, 1995).

Baker and Allen (1993b) fitted whole-day canopy Pn data to a rectangular hyperbola of the form of equation (4.1) for DAP 61 of the RICE II experiment. The parameters were 70.8 m mol/mol, 3.96 and -2.21, for K_m, Y_{max} and Y_i, respectively, with a relative Pn response ceiling of 1.75 at infinite [C]. The calculated percentage increase in response of whole-day canopy Pn at 660 vs. 330 m mol/mol was 36%.

Photosynthetic Acclimation to CO_2 at the Canopy Level

To test for acclimation of canopy photosynthetic capacity, Baker *et al.* (1990b) used rice that was grown at 160, 250, 330, 500, 660 and 900 m mol/mol CO_2. For half-day periods (mornings) on 62,63 and 64 DAP, common CO_2 setpoints of 160, 330 or 660 m mol/mol were imposed on each chamber on those respective days. Within each of these short-term CO_2 exposure comparisons, Baker *et al.* (1990b, 1996b) and Baker and Allen (1993a) showed that the short-term canopy net photosynthetic rate decreased with increasing long-term CO_2 treatment. The relative effects of the short-term CO_2

exposure were greatest for the lowest short-term CO_2 concentration. For example, when compared at a common short-term CO_2 exposure of 160 m mol/mol, the canopy photosynthetic rate of the chamber containing the 900 m mol/mol long-term CO_2 growth treatment was only about one-third of that of the 160 m mol/mol long-term CO_2 growth treatment. However, a large part of this apparent acclimation effect may be attributed to the greater respiration rates of the plants that had been grown under elevated CO_2 (Boote *et al.,* 1994).

Although Rowland-Bamford *et al.* (1991) reported decreases in both amount and activity of rubisco with increasing CO_2 in rice plants grown across the range of 160 to 900 m mol/mol, the leaf A/C. response curves gave no indication of a downward acclimation of photosynthetic capacity across the smaller range of 330 to 660 m mol/mol CO_2. One possibility would be that rubisco is not as limiting for photosynthesis under CO_2 enrichment as would be expected. More work is needed, under a range of CO_2 treatments, to explore the interaction effects of sink capacity, nitrogen nutrition, and other internal CO_2-fixation processes on photosynthetic behaviour and crop yield.

Comparison of rice canopy net photosynthetic rate (Pn) vs. long-term [CO_2] acclimation treatment for rice canopies grown at subambient (160, 250), ambient (330), and superambient (500, 660 and 900 m mol CO_2/mol air) [CO_2] treatments in 1987. The Pn estimates were obtained during a short-term [CO_2] change-over study where the [CO_2] was maintained during the morning hours in all six long-term [CO_2] treatments at 160, 330 and 660 m mol/mol on days 62, 63 and 64 days after planting, respectively. The Pn was estimated from linear regression equations of Pn vs. photosynthetic photon flux density (PPFD) with PPPD set to 1500 m mol photons/m^2/s. The vertical bars represent 95% confidence intervals. Adapted from Baker *et al.* (1990b, 1996b) and Allen *et al.*, (1995).

For soybean, leaf blade soluble protein expressed on a leaf blade area basis and percentage rubisco protein expressed on a leaf blade soluble protein basis for 34-day-old soybean plants grown under a wide range of CO_2 concentrations. Adapted from Campbell *et al.* (1988) and Baker and Allen (1994). For rice, leaf nitrogen content expressed on a leaf area basis and percentage rubisco protein expressed on a leaf soluble protein basis for 75-day-old rice plants grown under a wide range of CO_2 concentrations. Adapted from Rowland-Bamford *et al.* (1991) and Baker and Allen (1994)

Development Stages

In rice, two distinctly different rates of leaf appearance for vegetative and reproductive phases of growth were found (Baker *et al.,* 1990, 1996; Baker and Allen, 1994), as has been observed many times before (Yoshida, 1977; Vergara, 1980). Leaf appearance rates (leaves per day) are about twice as great in the vegetative phase as in the reproductive phase.

In the 'late' planted (23 June 1987) RICE II experiment, the number of mainstem leaves at panicle initiation and the final number of mainstem leaves decreased across CO_2 treatments of 160 to 500 m mol/mol and remained similar from 500 to 900 m mol/mol. Panicle initiation and boot stage occurred about 12 days earlier in the superambient CO_2 treatments compared with the 160 m mol/mol treatment (Baker *et al.,* 1990a; Baker and Allen, 1993b). Therefore, plant developmental rate was clearly accelerated with increasing CO_2 up to about 500 m mol/mol.

Temperature effect data were assembled from treatments ranging from 25/18/21 to 40/33/37°C (day/night/paddy water temperatures) in RICE I to RICE V experiments. Air temperature greatly influenced leaf appearance rate, developmental rate, and total growth duration, whereas, across the limited range of 330 to 660 m mol/mol the effects were comparatively small (Baker and Allen, 1993b; Baker *et al.,* 1992a,b, 1995). No consistent differences in phyllochron interval (days per leaf appearance) between 330 and 660 m mol/mol were observed, whereas phyllochron interval increased with increasing temperature across the range of 25/18/21 to 40/33/37°C, especially for the reproductive phase. In RICE III, IV and V experiments, anthesis ranged from 0 to 6 days earlier in the 660 vs. 330 m mol/mol treatments (Baker *et al.,* 1995). Also, the number of days to anthesis was shortened by approximately 10 days across the temperature treatment range from 25/18/21 to 34/27/31°C for the RICE IV experiment.

Growth and Yield

The yield and yield component data of the RICE I and RICE II experiments with CO_2 concentrations spanning 160 to 900 m mol/mol are given. A rectangular hyperbola (equation (4.1)) was fitted to the seed yield data. Values of K = 284 m mol/mol, Y_{max} =2.24 and Y_i= -0.13 were obtained for relative grain yield with asymptotic relative yield response ceiling (Y_{max} + Y_i) = 2.11 at infinite [C]. The calculated percentage increase in response at 660 vs. 330 m mol/mol was 44% for grain yield (Baker and Allen, 1993b).

Across a relatively wide temperature range from 25/18/21 to 37/30/34°C, Baker *et al.* (1992b) found a broad temperature optimum for biomass production in the mid-temperature ranges. Plants grown at 40/33/37°C were near the upper temperature limit for survival. High temperature spikelet sterility of rice is induced almost exclusively on the day of anthesis (Satake and Yoshida, 1978) when temperatures greater than 35 °C for more than one hour induce a high percentage of sterility (Yoshida, 1981). In the 40/33/37°C treatments, plants in the 330 m mol/mol CO_2 chamber died during internode elongation whereas plants in the 660 m mol/mol chamber produced small, abnormally shaped panicles that were sterile (Baker *et al.,* 1992a). Therefore, elevated CO_2 may slightly increase the maximum temperature at which rice plants can survive.

At both CO_2 levels, grain yield was highest in the 28/21/25°C treatment followed by a decline to zero yield in the 40/33/37°C treatment. The CO_2 enrichment from 330 to

660 m mol/mol increased yield by increasing the number of panicles per plant, whereas the number of filled grains per panicle and individual seed mass were less affected. Temperature effects on yield and yield components were highly significant. The number of panicles per plant increased while the number of filled grains per panicle decreased sharply with increasing temperature treatment. Individual seed mass was stable at moderate temperatures but tended to decline at temperature treatments above 34/27/31°C. Final above-ground biomass and harvest index were increased by CO_2 enrichment while harvest index declined sharply with increasing temperature. Notably, there were no significant CO_2 x temperature interaction effects on yield, yield components, or final above-ground biomass.

At each CO_2 concentration, polynomial regression equations were fitted to the rice seed yield (Y) vs. temperature (X) data. A third-degree polynomial [$Y = -239.226 + 25.557 * X — 0.848 * X^2 + 0.00896 * X^3$ ($R^2 = 0.91$, $S_{y.x} = 1.36$)] provided the best fit for the 660 m mol/mol CO_2 treatments (Baker and Allen, 1993a). However, a second-degree polynomial [$Y = -1.196 + 1.042 * X — 0.0278 * X^2$ ($R^2 = 0.91$, $S = 1.36$)] fits the data best for the 330 m mol/mol treatments. The effect of temperature on seed yield and final biomass, respectively, from a number of experiments based. While future increases in atmospheric CO_2 should benefit rice yields, large negative effects are likely if temperatures also rise.

Canopy Dark Respiration Rates

Plant respiration rates may be decreased by both short- and long-term exposure to high CO_2 concentrations (Bunce, 1990; Amthor, 1991); however Baker *et al.* (1992c) found that nighttime canopy dark respiration rates [R_d, m mol $(CO_2)/m^2$ (ground area)/s] increased for rice exposed to daytime CO_2 ranging from 160 to 900 m mol/mol. Similar to photosynthetic rates, R_d increased with CO_2 exposure from 160 to 500 m mol/mol but levelled off somewhat across the 500 to 900 m mol/mol range. The R_d of the ambient and superambient CO_2 treatments reached a broad maximum around 30 to 50 DAP whereas the broad maximum of the subambient treatments occurred later around 50 to 70 DAP. The maximum values of R_d were about 6, 8, and 9 m mol/m²/s for the 160, 250 and 330 m mol/mol CO_2 treatments and about 11 to 12 m mol/m²/s for the three superambient CO_2 treatments.

Specific respiration rate [R_{dw}, m mol (CO_2)/s/kg (total above-ground dry matter)] decreased exponentially with DAP at all CO_2 exposures and was higher in the subambient (160 and 250 m mol/mol) than in the ambient (330 m mol/mol) and superambient (500, 660 and 900 m mol/mol treatments). At each CO_2 exposure level, the patterns of R_{dw} with DAP were very similar to the patterns of plant tissue nitrogen concentration with DAP (Baker *et al.,* 1992c). Furthermore, R_{dw} was linearly related to total above-ground plant tissue nitrogen concentration [mg (N)/g (DW)] across the range of CO_2 exposures and the six dates of plant sampling ($R_{dw} = -13.0 + 0.952 * [N]$; r2=0.9, P=0.01). Baker *et al.* (1992c) concluded that the CO_2 treatments affected R_{dw} mainly by altering the protein content of the plant tissue. Another explanation is that elevated CO_2 increased the amount of structural and non-structural carbohydrates in the plant tissues, so that a larger proportion of dry matter was sequestered in non-protein materials. This study showed no respiration acclimation to long-term CO_2 enrichment (indirect acclimation) of rice that could not be explained by nitrogen concentration of the plant tissues.

Evapotranspiration and Water-use Efficiency

Both plant transpiration and direct evaporation from the floodwater surface contribute to water use (evapotranspiration, ET) of rice growing in the SPAR chambers. Diurnal trends of ET followed diurnal patterns of solar irradiance (Baker *et al.*, 1990b; Allen *et al.*, 1995). After canopy closure, maximum ET rates and total daytime water losses were about 35 and 30% greater, respectively, from rice grown at 160 m mol/mol compared to rice grown at 900 m mol/mol CO_2. The ET rates were similar when all chambers were exposed for one-half day to the same CO_2 concentration (data not shown, Baker *et al.*, 1990b), which demonstrates the effect of the exposure level of CO_2 on stomatal control of transpiration.

Temperature also has a large effect on diurnal ET rates (Baker and Allen, 1993a; Allen *et al.*, 1995) and daytime water use, mediated primarily through vapour pressure deficit of the air. Solar irradiance also has a large effect, probably through both the energy inputs to the canopy and through the stomatal opening response to light (directly or indirectly). Midday maximum ET rates were about 75 and 35% higher for rice grown at 40/33/37°C and 34/27/31°C, respectively, compared to rice grown at 28/21/25°C (Baker and Allen, 1993a; Allen *et al.*, 1995).

Water-use Efficiency: Effects of CO_2

Daytime totals of CO_2 uptake, ET, and calculations of WUE vs. CO_2 treatment. Stomatal conductance decreases with increasing CO_2 concentration which can cause a reduction of both leaf and whole canopy transpiration. However, CO_2 enrichment may also increase canopy leaf surface area for transpiration, thereby offsetting some of the water savings (Jones *et al.*, 1985b; Allen *et al.*, 1985). The leaf area index of the RICE II experiment ranged from 7.6 to 10.8 across the CO_2 treatments from 160 to 900 m mol/mol. Calculations of WUE increased with increasing CO_2 due to the decline in ET across this CO_2 range and the increase in Pn with CO_2 up to the 500 m mol/mol treatment.

Water-use Efficiency: Effects of Temperature

In the 28°C air temperature treatment, CO_2 enrichment from 330 to 660 m mol/mol resulted in 30% increase in daytime total CO_2 uptake and 22% decrease in ET with a concomitant increase of WUE from 1.42 to 2.37 m mol (CO_2)/mmol (H_2O), an increase of 67%. Daytime ET was more than doubled from the 25 to the 37°C daytime temperature treatment. WUE declined from the 28 to the 37°C daytime treatment due to the sharp increase in ET and the relatively stable Pn across this temperature range. If WUE were based on grain yield, it would decrease drastically with increasing temperature not only because of increasing ET, but also because of sharply decreasing seed production.

Rising atmospheric CO_2 is likely to benefit rice production by increasing photosynthesis, growth and grain yield while reducing water use and increasing WUE. In warm areas of the world, however, possible future global warming may result in substantial yield decreases because of the sensitivity of flowering and seed set to high temperatures and the possibility of water shortages that may result from increased evapotranspiration.

Summary of Comprehensive Reviews

Several recent symposia proceedings and reviews leave little doubt that crop plants can respond well to elevated CO_2 (Rozema *et al.*, 1993; Woodwell and Mackenzie, 1995; Wittwer, 1995). Poorter (1993) compiled information from 156 plant species and found that doubling CO_2 provided an average growth increase of 37%. The distribution of weight ratios of CO_2-enriched and control plants. Poorter's compilation showed a 41 and a 22% increase for C_3 and C_4 plants, respectively. As a group, C_3 herbaceous crop plants responded more than wild herbaceous species (58 vs. 35%). Furthermore, the fast growing wild species responded more strongly than slow-growing wild species (54 vs. 23%).

Poorter (1993) imposed two restrictions on his compilations that may have led to larger than expected responses to elevated CO_2. Firstly, plants grown in competition were not included. Secondly, only vegetative stages of plants were compared since compiled data were selected prior to flowering.

A number of studies have shown that vegetative growth responses may be greater than reproductive (seed yield) responses. Therefore, the compiled data of Poorter may give an impression of greater response than would be observed throughout the life cycle. Secondly, crops in field conditions usually are grown in dense populations where they compete for space and light. Under more realistic field conditions, crop plants are likely to respond as a community rather than individual plants, wherein light (solar radiation) becomes a limiting factor for growth. Under these conditions, elevated CO_2 cannot promote horizontal expansion and greater light capture. Although the actual field responses may be less, the CO_2 fertilization effect is clearly well-established.

The CO_2 fertilization effect for forest species has also become firmly established. Wullschleger *et al.* (1994) estimated the biotic growth factor for 58 controller-exposure studies of forest tree species which included 398 observations. Their frequency distribution of relative growth response of trees grown at elevated CO_2 vs. ambient CO_2 is given.

They also found a mean response ratio of 1.32. Of the 398 observations, 51 showed a relative growth response less than 1, and 31 showed responses greater than 2. However, under competitive conditions, tree response may be much less (Bazzaz *et al.,* 1995).

Kimball *et al.* (1993) discussed the data of Idso *et al.* (1987) which related the growth modification factor (or biomass growth modification ratio) caused by a 300 m mol/mol increase in CO_2 concentration above ambient. Their data are reproduced. Under Phoenix, Arizona, conditions, the growth modification factor across a mean temperature range of 12 to 34 °C was rather linear. Thus, the *vegetative* response to CO_2 should be enhanced at increased temperatures. As discussed before, this growth modification factor may not apply for reproductive growth responses (seed yield) of crops (e.g., rice).

Crop Modelling: Predictions for the Future

Modelling Crop Responses to Co_2 and Climate Changes

Peart *et al.* (1989) and Curry *et al.* (1990a,b, 1995) predicted growth and yield responses of soybean and maize to doubled-CO_2 climate change scenarios of the southeastern USA. Their simulations used 30 years of baseline weather data (1951-1980) from 19 sites in 11 states. Predicted climate changes of the Goddard Institute for Space Studies (GISS) model (Hansen *et al.,* 1988) and the Geophysical Fluid Dynamics Laboratory (GFDL) model (Manabe and Wetherald, 1987) were used to change temperatures, precipitation and solar radiation, month-by-month, for the 30 years of baseline data at each site. These modified baseline weather data provided GISS and GFDL climatic change scenarios (Smith and Tirpak, 1989). At each site, prevailing cropped soil types, planting dates, and cultivars were used in the simulations.

Baseline, GISS and GFDL scenario temperature and rainfall data that were used by Peart *et al.* (1989) were condensed for two representative locations; Columbia, South Carolina, and Memphis, Tennessee. Crop yield responses to climate change were simulated under four conditions: with or without direct CO_2 fertilization effects, and under rainfed or optimum irrigation culture. The Penman-Monteith equation, which contains a term for canopy conductance, was used to compute the effects of elevated CO_2 on canopy transpiration. Simulations were run for doubled-CO_2 conditions with a crop photosynthetic enhancement factor of 1.35 for soybean (a C_3 plant) and 1.10 for maize (a C_4 plant). No simulations were run with CO_2 fertilization effects only (without climate change effects). Simulated soybean seed yields were averaged over 30 years and 19 sites. Under rainfed conditions with climate change effects only, simulated soybean yields under the GFDL scenario were reduced 71 % compared to the baseline climate, whereas the average yields were reduced only 23% under the GISS scenario. Yields under the GFDL scenario were severely impacted because of the rainfall reductions predicted by this model.

Under optimum irrigation conditions, average yields under both the GISS and GFDL scenarios were reduced 18 to 19%. However, in spite of higher temperatures, the irrigated yields under the GISS and GFDL scenarios were about 25% greater than the baseline climate scenario without irrigation.

When CO_2 fertilization effects were included with climate change effects, simulated average yields of the GISS climate scenario were increased 11% under rainfed conditions, whereas yields under the GFDL climate scenario were still decreased (-52%). Under optimum irrigation with CO_2 fertilization effects, yields under both GISS and GFDL scenarios were increased 13 to 14%.

Yields of maize were simulated for doubled-CO_2 fertilization effects at only four weather station locations (Charlotte, North Carolina; Macon, Georgia; Memphis, Tennessee; Meridian, Mississippi). For climatic change effects alone, predicted maize yields declined only 6% in the GISS scenario, but declined 73% in the GFDL scenario. Although irrigation increased predicted crop yields, the GISS and GFDL climate scenarios gave yield decreases of 18 and 27%, respectively, with respect to the irrigated baseline weather. The yield reduction of the GFDL scenario with respect to the GISS scenario was 10%, attributable to slightly higher temperatures. Including the CO_2 fertilization effects with climatic change scenarios had little effect on the predicted yields of maize because it is a C_4 plant.

In the Great Lakes and Corn Belt Area, simulations by Ritchie *et al.* (1989) showed that higher temperature would have the greatest effect of the climate change factors on predicted yields of soybean and maize, mainly through decreases in the crop life cycle. For the southern part of this region, yield reductions were greater for the GFDL scenario. Predicted yields increased for the northernmost stations because temperatures and growing season duration became more favourable. Average irrigation water requirements in this region increased about 90% for the GISS and GFDL scenarios.

Rosenzweig (1989) modelled maize and wheat yields in the Great Plains under GISS and GFDL scenarios. Yields decreased in the Southern Great Plains because higher temperatures shortened the life cycle of the crops. Where precipitation was predicted to decrease, irrigation requirements increased. In a separate modelling study, Allen and Gichuki (1989) predicted a 15% increased requirement for irrigation for this region, with greater requirements for alfalfa because its growing season was increased, and lower requirements for maize and winter wheat because their growing seasons were decreased. The CO_2 fertilization effect offset the adverse effects of climate change at some locations of Ritchie *et al.* (1989) and Rosenzweig (1989).

Dudek (1989) predicted productivity changes of several Californian vegetable, fruit and nut crops in response to GISS and GFDL scenarios for doubled CO_2. Without CO_2 fertilization effects, statewide average yield changes, depending on the crop, would be

-8 to -34% for the GISS scenario and -6 to -31 % for the GFDL scenario. With CO_2 fertilization effects, predicted statewide yield changes ranged from about +17 to -12% for the GISS scenario and about +21 to -8% for the GFDL scenario.

Dudek (1989) used a California Agriculture and Resource Model (CARM) to further predict economic and market impact of the productivity changes. Production declined, in general, under the scenarios of climate change without CO_2 fertilization effects; however, commodity prices generally increased. Under the CO_2 fertilization plus climate change effects scenarios, predicted impacts on commodity prices were much less.

The effect of temperature on the phenology of crop plants plays a critical role. One crucial need is for more detailed research on responses of plants to temperature, and temperature x CO_2 interactions, as inputs to crop models. More modelling studies are also needed on different planting dates as an adaptive strategy. Cultivars need to be designed for future climatic conditions (Hall and Allen, 1993). Factors that should be considered are: extension of the grain-filling period and perhaps shortening the duration of vegetative growth (which would also improve harvest index); ability to flower and set seed at higher temperatures; photoperiod and thermoperiod adaptive interactions; selection for positive photosynthetic acclimation where negative photosynthetic acclimation has been observed; and capability of utilizing photoassimilates (storage carbohydrates) more effectively. These factors need to be integrated into whole-plant physiology under real-world conditions.

Adaptations and Evapotranspiration Requirements

The simulated crop yield responses to climatic changes provided by Peart *et al.* (1989) and Curry *et al.* (1990a,b, 1995) manifest two main points: (a) the serious adverse impact of inadequate rainfall scenarios on crop production coupled with rising temperature scenarios, and (b) the importance of beneficial CO_2 fertilization effects in the face of elevated temperatures. However, the climate change for an effective doubling of CO_2 may occur at CO_2 concentrations less than those used in this simulation, if radiatively active trace gases other than CO_2 play a large role in the greenhouse effect. In that case, the direct CO_2 effects would be somewhat lower than shown in the example, for an equivalent climate change. All of the simulations assumed that climatic changes would occur simultaneously with increasing concentrations of CO_2 and other trace greenhouse-effect gases. If global warming lags the increases of atmospheric CO_2, then some beneficial effects of CO_2 fertilization are likely to occur before the full impact of climate change is manifested. However, Broecker (1987) and others caution that climate changes have not always been gradual during interglacial periods of the Pleistocene Era. There is clear evidence of relatively rapid climatic oscillations in the northern hemisphere during the previous interglacial period 110000 to 140000 years before present based on Greenland Ice-core Project (GRIP) records

(Anklin *et al.,* 1993). These oscillations produced cold periods that were as severe as the proceeding glacial period.

Much of the reduction in soybean yields reported by Peart *et al.* (1989) and Curry *et al.* (1990a,b, 1995) was caused by decreases in the length of the grain-filling period under higher temperatures. Changes in management practices, such as changing planting dates or selection of other cultivars, may help to prevent some of the potential reductions in yield. In the future, plant breeders may need to adapt combinations of temperature tolerance and photoperiod responses into new germplasm. In situations where non-structural carbohydrates accumulate as a CO_2 fertilization effect response, new germplasm needs to be developed that can make better use of the photoassimilate source.

Irrigation is not likely to be a panacea for climate change. Predicted irrigation requirements for soybean in the southeastern USA were increased 33 and 134% under the GISS and GFDL scenarios, respectively, based on simulations of Peart *et al.* (1989) and Curry *et al.* (1990a,b). However, under the GFDL scenario, water resources would become scarce, and may not be readily available for crops. Some areas of the USA may have to adapt by irrigating less land area. Increasing temperatures and decreasing precipitation for the USA as predicted by the GFDL model would have a serious negative impact overall on agricultural productivity although producers in favourable regions may benefit from scarcity-mediated higher prices (Adams *et al.,* 1990).

Assessing International Impacts of Rising Co_2 and Climate Change

Several assessments have been conducted on the impacts of rising CO_2 and global climatic changes on crop production patterns and economics responses within national or regional zones (Adams *et al.,* 1990; Crosson, 1993; Parry *et al.,* 1988; Smith and Tirpak, 1989) and in various countries around the world (Rosenzweig and Iglesias, 1994; Rosenzweig *et al.,* 1995). Furthermore, progress has been made on predicting impacts of climate change using world trade models (Rosenzweig and Parry, 1993, 1994).

The studies conducted by Rosenzweig and colleagues used existing crop models and climate change scenarios from three GCMs (GISS, GFDL and United Kingdom Meteorological Office Model (UKMO), Wilson and Mitchell, 1987) to predict changes in crop yields for a number of countries around the world. The doubled-CO_2 climate change scenarios (temperature, rainfall and evaporation changes) were based on the climate change potential expected from increases of all greenhouse-effect gases. These climate changes are expected to occur well before CO_2 concentration has actually doubled. Therefore, the CO_2 levels for fertilization effects were estimated to be 555 m mol/mol rather than 660 m mol/mol. Thus, the CO_2 fertilization effect photosynthetic ratios for four crops (soybean, wheat, rice and maize) were taken as 1.21, 1.17, 1.17 and 1.06, respectively.

The impact of climate change scenarios was more severe in the tropical latitudes than in the mid- or high-latitudes. For example, averaged over all three GCM scenarios, wheat production changes predicted in Brazil, India, China and Canada were -47, -43, -11 and -20%, respectively, without CO_2 effects, and -28, -13, +8 and +16% with CO_2 fertilization effects (calculated from data of Rosenzweig and Parry, 1993; Rosenzweig and Iglesias, 1994).

Simulations of soybean yields throughout the USA were a part of this international study (Curry *et al.*, 1995). They found that aggregated yields were 2.42, 2.80, 2.37 and 1.31 Mg/ha for the BASE, GISS, GFDL and UKMO scenarios, respectively. Temperatures were about 4.0°C higher than BASE for the GISS and GFDL scenarios, but were about 5.2°C higher for the UKMO model.

Refinements and improvements in prediction methodology will continue, but these assessments provide the best currently available insights into the CO_2 fertilization and climate change effects on global crop productivity.

Elevated CO_2 increases the size and dry weight of most C_3 plants and plant components. Relatively more photoassimilate is partitioned into structural components (stems and petioles) during vegetative development in order to support the light-harvesting apparatus (leaves). The harvest index tends to decrease with increasing CO_2 concentration and temperature. Selection of plants that could partition more photoassimilates to reproductive growth should be a goal for future research. As more is learned about the effects of anticipated climate changes on crops, more effort should be directed to exploring biological adaptations and management systems for reducing these impacts on agriculture and humanity. Whether regional climates become drier or wetter with global warming remains to be seen.

Multifunctional Dimensions of Ecologically-based Agriculture in Latin America : A Case Study

Today in Latin America there are still regions with microcosms of traditional farming systems, (i.e., in Mesoamerica, the Andean region, and the Amazon Basin) that have emerged over centuries of cultural and biological evolution and that based on locally available resources and the cultivation of a diversity of crops and varieties in time and space, have allowed traditional farmers to maximize harvest security and the multiple use of the landscape with limited environmental impact. Agro-biodiverse traditional agroecosystems represent a strategy which ensures diverse diets and income sources, stable production, minimum risk, efficient use of land resources, and enhanced ecological integrity. This legacy of traditional agriculture demonstrate that the combination of stable and diverse production, internally generated and maintainable inputs, favorable energy input/output ratios, and articulation with both subsistence and market needs, comprises an effective approach to achieve food security, income generation, and environmental conservation. Traditional approaches represent multiple use strategies that enhance the multifunctional nature of agriculture, an important feature for the health of rural regions in the next century.

Introduction

Agriculture is a process of artificialization of nature. In general, modern agriculture has implied the simplification of the structure of the environment over vast areas, replacing nature's diversity with a small number of cultivated plants and domesticated animals. In fact, the world's agricultural landscapes are planted with only some 12 species of grain crops, 23 vegetable crop species, and about 35 fruit and nut type species, that is no more than 70 plant species spread over approximately 1,440 million hectares of presently cultivated land in the world, a sharp contrast with the diversity of plant species found within one hectare of a tropical rainforest which typically contains over 100 species of trees (Thrupp 1998).

But not all forms of agriculture have followed the classic path of artificialization and intensification. In Latin America, systems range from "low intensity" long-fallow swidden to "high intensity" permanent cultivation wherein large areas have been greatly modified from their natural state and are dominated by monocultures. In commercial agricultural areas, natural habitats are lost through expansion of agricultural production, especially of cattle, sugarcane, cotton, soybean, coffee, and (recently) non-traditional export crops. Highly capitalized farms tend either to be on high-quality lands where profitability is contingent on low wages and large landholdings. By contrast, farms of resource-poor peasants tend to be on ecologically marginal lands or on lands recently opened to agriculture. Thus, impoverished farmers lack access to good farmland and capital and are forced by necessity onto remnants of natural areas, which generally occur on steep slopes, along rivers, and in other fragile environments such as forest margins.

In the midst of these extreme types of agriculture, there are, in the region microcosms of traditional farming systems, (i.e., in Mesoamerica, the Andean region, and the Amazon Basin) that have emerged over centuries of cultural and biological evolution and represent accumulated experiences of peasants interacting with the environment without access to external inputs, capital, or scientific knowledge (Chang, 1977; Wilken, 1987). Using inventive self-reliance, experiential knowledge, and locally available resources, indigenous farmers have often developed farming systems with sustained yields (Harwood, 1979; Reinjtes et al., 1992). These agroecosystems, based on the cultivation of a diversity of crops and varieties in time and space, have allowed traditional farmers to maximize harvest security under low levels of technology and with limited environmental impact (Clawson, 1985). There are also several examples of grass-roots rural development programs in Latin America aimed at the maintenance and/or enhancement of biodiversity in traditional agroecosystems, and which represent a strategy which ensures diverse diets and income sources, stable production, minimum risk, efficient use of land resources, and enhanced ecological integrity (Altieri, 1995; Pretty, 1995).

Increasingly, evidence emerging from analysis of traditional agriculture and NGO-led agroecological projects, shows that the combination of stable and diverse production, internally generated and maintainable inputs, favorable energy input/output ratios, and articulation with both subsistence and market needs, comprises an effective approach to achieve food security, income generation, and environmental conservation (Pretty, 1997; Altieri et al., 1998). As it will be argued in this paper, these approaches represent multiple use strategies that enhance the multifunctional nature of agriculture.

The Multifunctional Nature of Traditional Agriculture

Despite the increasing industrialization of agriculture, the great majority of the farmers in the developing world are peasants, or small producers, who still farm the valleys and slopes of rural landscapes with traditional and subsistence methods. It is estimated that in Latin America there are about 16 million peasant units occupying close to 160 million hectares and involving 75 million people, representing two-thirds of the regions total rural population (Ortega 1986).

Many of these agroecosystems are small-scale, geographically discontinuous, and located on a multitude of slopes, aspects, microclimates, elevational zones, and soil types. They also are surrounded by many different vegetation associations,. The combinations of diverse physical factors therefore are numerous and are reflected in the diverse cropping patterns chosen by farmers to exploit site-specific characteristics. Many of the systems are surrounded by physical barriers (e.g., forests, rivers, mountains) and therefore are relatively isolated from other areas where the same crops are grown in large scale. Descriptions of the species and structural diversity and management of these traditional systems are discussed elsewhere (Alcorn, 1984; Altieri et al.,1987; Chang, 1977; Clawson, 1988; Denevan, 1995; Francis, 1986; Toledo et al. 1985).

In many areas, traditional farmers have developed and/or inherited complex farming systems, adapted to the local conditions helping them to sustainably manage harsh environments and to meet their subsistence needs, without depending on mechanization, chemical fertilizers, pesticides or other technologies of modern agricultural science (Altieri, 1995). According to Toledo (1995), indigenous farmers in the hot and humid tropical regions of Latin America tend to combine various production systems as part of a typical household resource management scheme:

1. The milpa system, which may constitute a system of polyuculture including up to 20-25 agricultural and forest species (annual and perennial) and is focused on the cultivation of maize, but in many occasions is combined and even substituted by agricultural market-oriented products (hot pepper, rice, sesame seeds, sugarcane, beans, etc.);

2. The extraction of products from the primary or secondary rainforests of different ages undergoing the succession process;
3. The manipulation of forest-unit sequences at different stages of anthropic disturbance, from which certain marketable products (mainly coffee, vanilla, and cocoa) are obtained;
4. The management of home gardens, which are agroforestry systems located next or close to households.

The main features underlying the sustainability of these multiple use peasant systems are (Marten, 1986; Reinjtes et al. 1992):

- Farms are small in size with continuous production serving subsistence and market demands
- Maximum and effective use of local resources and low dependence on off-farm inputs
- High net energy yield because energy inputs are relatively low
- Labor is skilled and complementary, drawn largely from the household or community relations. Dependency on traction and manual labor shows favorable energy input/output ratios
- Heavy emphasis is on recycling of nutrients and materials
- Building on natural ecological processes (e.g., succession) rather than struggling against them
- Diversified farm systems based on several cropping systems, featuring mixtures of crops, and crops with varietal and other genetic variability.

A salient feature of traditional farming systems is their degree of plant diversity, generally in the form of polycultures and/or agroforestry patterns (Clawson, 1985). This peasant strategy of minimizing risk by planting several species and varieties of crops stabilizes yields over the long term, promotes diet diversity, and maximizes returns under low levels of technology and limited resources (Richards, 1985). Traditional multiple cropping systems provide as much as 20 percent of the world food supply (Francis, 1986). Polycultures constitute at least 80 percent of the cultivated area of West Africa, while much of the production of staple crops in the Latin American tropics occurs in polycultures. Polycultures produce more combined yield in a given area than could be obtained from monocultures of the component species. Most traditional polycultures exhibit LER values greater than 1.5. Moreover, yield variability of cereal/legume polycultures are much lower than for monocultures of the components.

Many traditional agroecosystems are located in centers of crop diversity, thus containing populations of variable and adapted land races as well as wild and weedy relatives of crops. It is estimated that throughout the Third World more than 3,000 native grains, roots, fruits and other food plants can still be found (Altieri and Merrick, 1987). Thus traditional agroecosystems essentially constitute in-situ repositories of genetic diversity (Altieri *et al.* 1987). Descriptions abound regarding systems in which tropical farmers plant multiple varieties of each crop, providing both intraspecific and interspecific diversity, thus enhancing harvest security. For example, in the Andes, farmers cultivate as many as 50 potato varieties in their fields (Brush *et al.* 1981). Similarly, in Thailand and Indonesia, farmers maintain a diversity of rice varieties in their paddies which are adapted to a wide range of environmental conditions, and regularly exchange seeds with neighbors (Grigg, 1974).

Tropical agroecosystems composed of agricultural and fallow fields, complex home gardens, and agroforestry plots, commonly contain well over 100 plant species per field and provide construction materials, firewood, tools, medicines, livestock feed, and human food. Home gardens in Mexico and the Amazon display highly efficient forms of land use, incorporating a variety of crops with different growth habits.

Small areas around peasant households commonly average 80-125 useful plant species, many for food and medicinal use (Toledo *et al.* 1985; Alcorn, 1984). Perennials such as fruit trees are a conspicuous feature of most homegardens (Marten, 1986). In some of the more humid areas, there are so many different kinds of trees and field cops in the homegardens, and they are growing in such abundance that it looks more like a tropical forest than a garden (Clarke and Thaman, 1993). Most diverse homegardens are in reality a collection of domesticated and semi-domesticated plants with a variety of uses including food, fuel, construction materials, herbal medicine, ornamentation, and shade. Homegardens are often in continuous production throughout the year and lend themselves to intensive care because they are so conveniently close to the house. They can be fertilized with kitchen wastes, receive supplementary irrigation with well water, and be attended by women and children in their spare time.

The Interface of Traditional Agroecosystems and Natural Areas

Most of the above studies of traditional agriculture have focused on the productive units where crops are grown. This limited view of the peasant agroecosystem ignores the fact that many peasants utilize, maintain, and preserve, within or adjacent to their properties, areas of natural ecosystems (forests, hillsides, lakes, grasslands, streamways, swamps etc.) that contribute valuable food supplements, construction materials, medicines, organic fertilizes, fuels, religious items, etc. (Toledo et al. 1985). In fact, the crop-production units and adjacent ecosystems constitute a continuum where plant gathering, fishing, and crop production are actively produced.

For many peasant societies, agriculture is considered a part of a bigger system of land use. For example, the P'urhepecha Indians who live in the region of lake Patzcuaro in Michoacan, Mexico, in addition to agriculture, gathering is part of a complex subsistence pattern based on multiple uses of their natural resources (Caballero and Mapes, 1985). These people use more than 224 species of wild native and naturalized vascular plants for dietary, medicinal, household, and fuel needs. Similarly, the Jicaque Indians of central Honduras, who live on the Montana de la Flor reservation, use over 45 plant species from the pine-oak forest, riverine habitat, or dooryard as foods, medicines, fuel, etc. Like their mestizo neighbors, the Jicaque grow corn using slash and burn techniques. The cultivated fields are widely spaced throughout the forest and in travelling from one field to the next, the Jicaque usually collect wild plant food along the way to be added to the cooking pots of the family's compound (Lentz, 1986).

Agriculture- natural ecosystem interfaces are of key significance as it has been shown that farmers accrue general ecological services from natural vegetation growing near their properties. For example, in many highland regions of Central America, the indigenous flora of the higher forests, not only provide valuable native plants for commercial and subsistence products, but also serve as natural barriers to the lowland agricultural crops against the spread of plant diseases and insect pests. Also, clearing comparatively small agricultural plots in a matrix of secondary forest vegetation permits easy emigration of natural enemies of insect pests from the surrounding jungle (Altieri, 1984).

In western Guatemala, small farms depend on nearby forests to manage marginal infertile soils. Leaf litter is carried from nearby forests and spread each year over intensively cropped vegetable plots to improve tilth and water retention. Litter is raked up, placed in bags or nets, and carried to fields by men or horses, or from more distant sources, by trucks. After spreading, the leaf litter is worked into the soil with a broad hoe. In some cases, litter is first placed beneath stable animals, and then, after a week or so the rich mixture of pulverized leaves, manure, and urine is spread over the fields and turned under. Although the quantities applied vary, farmers in Almolonga, Zunil, and Quezaltenango apply as much as 40 metric tons of litter/ha. each year. Rough calculations made in mixed pine-oak stands indicate that one hectare of cropped land requires the litter production from 10 ha. of regularly harvested forest, or less, if harvesting is sporadic (Wilken, 1987).

A Case Study of a Multifunctional Traditional Farming System

The study conducted in a Totonaca native community of the Papantla region in the state of Veracruz illustrates of a case multiple use peasant management strategy of hot and humid tropical ecosystems. The community entails 166 households totaling a population of 877 and sharing a 15-17 hectare territory. Most households (72%) have between 7 and 9 hectares, while only 9 % own more than nine hectares and 19% less than seven hectares. Most of these households also handle from 3 to 9 ecogeographic or landscape units as resources for production where they implement the multiple-use strategy. The main units that each family manages during production are: milpa (maize fields), pasture ground, home gardens, rainforest for vanilla production, rainforest to extract wood and other products, and cash crop areas.

Using almost exclusively its own physical energy (with scant, almost inexistent use of chemical fertilizers), making little use of outside inputs, and relying on family or community labor, the productive units of this native community are self-sufficient in terms of food, they are energy efficient, they do not generate waste, and they sustain a high level of agrobiodiversity (with 355 species of plants, animals, and fungi). To this should be added the fact the community succeeds in being economically profitable as a result of selling maize, beef, milk, vegetable, fruits, vanilla, brown sugar, palm leaves and other products (Toledo, 1995).

The Nature and Function of Biodiversity in Agriculture

Today, scientists worldwide are increasingly starting to recognize the role and significance of biodiversity in the functioning of agricultural systems (Swift et al., 1996). Research suggests that whereas in natural ecosystems the internal regulation of function is substantially a product of plant biodiversity through flows of energy and nutrients and through biological synergisms, this form of control is progressively lost under agricultural intensification and simplification, so that monocultures, in order to function, must be predominantly subsidized by chemical inputs (Swift et. al. 1996). Commercial seed-bed preparation and mechanized planting replace natural methods of seed dispersal; chemical pesticides replace natural controls on populations of weeds, insects, and pathogens; and genetic manipulation replaces natural processes of plant evolution and selection. Even decomposition is altered since plant growth is harvested and soil fertility maintained, not through nutrient recycling, but with fertilizers.

One of the most important reasons for maintaining and/or encouraging natural biodiversity is that it performs a variety of ecological services (Altieri, 1991). In natural ecosystems, the vegetative cover of a forest or grassland prevents soil erosion, replenishes ground water, and controls flooding by enhancing infiltration and reducing water runoff. In agricultural systems, biodiversity performs ecosystem services beyond production of food, fiber, fuel, and income. Examples include, recycling of nutrients, control of local microclimate, regulation of local hydrological processes, regulation of the abundance of undesirable organisms, and detoxification of noxious chemicals. These renewal processes and ecosystem services are largely biological, therefore their persistence depends upon maintenance of biological diversity. When these natural services are lost due to biological simplification, the economic and environmental costs can be quite significant. Economically in agriculture, the burdens include the need to supply crops with costly external inputs, since agroecosystems deprived of basic regulating functional components lack the capacity to sponsor their own soil fertility and pest regulation. As functional biodiversity decreases, the requirement for higher management intensity increases, thus monocultures must be subsidized with external inputs. Often, the costs involve a reduction in the quality of the food produced and of rural life in general due to decreased soil, water, and food quality when erosion and pesticide and/or nitrate contamination occurs (Altieri, 1995).

Biodiversity refers to all species of plants, animals and microorganisms existing and interacting within an ecosystem. In agroecosystems, pollinators, natural enemies, earthworms, and soil microorganisms are all key biodiversity components that play important ecological roles thus mediating processes such as genetic introgression, natural control, nutrient cycling, decomposition, etc. The type and abundance of biodiversity in agriculture will differ across agroecosystems which differ in age, diversity, structure, and management. In fact, there is great variability in basic ecological and agronomic patterns among the various dominant agroecosystems. In general, the degree of biodiversity in agroecosystems depends on four main characteristics of the agroecosystems (Southwood and Way, 1970):

1. the diversity of vegetation within and around the agroecosystem
2. the permanence of the various crops within the agroecosystem
3. the intensity of management
4. the extent of the isolation of the agroecosystem from natural vegetation.

In general, agroecosystems that are more diverse, more permanent, isolated, and managed with low input technology (i.e. agroforestry systems, traditional polycultures) take fuller advantage of work done by ecological processes associated with higher biodiversity than highly simplified, input-driven and disturbed systems (i.e. modern row crops and vegetable monocultures and fruit orchards) (Altieri, 1995).

All agroecosystems are dynamic and subject to different levels of management so that the crop arrangements in time and space are continually changing in the face of biological, cultural, socio-economic, and environmental factors. Such landscape variations determine the degree of spatial and temporal heterogeneity characteristic of agricultural regions, which in turn conditions the type of biodiversity present.

According to Vandermeer and Perfecto (1995), two distinct components of biodiversity can be recognized in agroecosystems. The first component, planned biodiversity, is the biodiversity associated with the crops and livestock purposely included in the agroecosystem by the farmer, and which will vary depending on management inputs and crops spatial/temporal arrangements. The second component, associated biodiversity, includes all soil flora and fauna, herbivores, carnivores, decomposers, etc., that colonize the agroecosystem from surrounding environments and that will thrive in the agroecosystem depending on its management and structure. The relationship of both biodiversity components. Planned biodiversity has a direct function, as illustrated by the bold arrow connecting the planned biodiversity box with the ecosystem function box. Associated biodiversity also has a function, but it is mediated through planned biodiversity. For example, the trees in an agroforestry system create shade, which makes it possible to grow only sun-tolerant crops. So the direct function of this second species (the trees) is to create shade. Yet along with the trees might come small wasps that seek out the nectar in the tree's flowers. These wasps may in turn be the natural parasitoids of pests that normally attack the crops. The wasps are part of the associated biodiversity. The trees, then, create shade (direct function) and attract wasps (indirect function) (Vandermeer and Perfecto, 1995).

The key is to identify the type of biodiversity that is desirable to maintain and/or enhance in order to carry out ecological services, and then to determine the best practices that will encourage the desired biodiversity components. There are many agricultural practices that have the potential to enhance functional biodiversity, and others that negatively affect it. The idea is to apply the best management practices in order to enhance and/or regenerate the kind of biodiversity that can subsidize the sustainability of agroecosystems by providing ecological services such as biological pest control, nutrient cycling, water and soil conservation, etc.

The Link Between Agrobiodiversity and Multifunctionality

When agricultural development takes place in a natural environment, it tends to result in a heterogeneous mosaic of varying types of habitat patches spread across the landscape. The bulk of the land may be intensely managed and frequently disturbed for the purposes of agricultural production, but certain parts (wetlands, riparian corridors, hillsides) may be left in a relatively natural condition, and other parts (borders and strips between fields, roadsides, and adjacent natural areas) may occasionally be disturbed but not intensely managed. In addition, natural ecosystems may surround or border areas in which agricultural production dominates (Gliessman, 1998).

The heterogeneity of the agricultural landscape varies greatly by region. In some parts of Latin America, where commercial, export agriculture predominates, the heavy use of agricultural chemicals, mechanical technology, narrow genetic lines, and irrigation over large areas have made the landscape relatively homogenous. In such areas, the agricultural landscape is made up mostly of large areas of single crop agricultural production. The expansion of such agricultural landscapes disrupts natural areas in three important ways. First, natural ecosystems become fragmented and important ecological linkages may be changed or uncoupled. For example, the conversion of uplands from native grasslands or deciduous forest to cotton will profoundly affect the nutrient and pesticide inputs into any adjacent wetlands. Second, the fragmentation increases boundary phenomena by increasing the proportion of area that is near a boundary. This results in an exacerbation of the impacts from adjacent agriculture. Third, the absolute loss of natural areas generally means that the remaining patches are increasingly more distant from each other. Thus each remnant takes on more and more the properties of oceanic islands in the sense that source areas for recolonization are often very distant. Thus, local extinction events for both species and genes are unlikely to be balanced by recolonization or gene flow. Unlike real islands, remnant patches of natural ecosystems are highly vulnerable to invasion by weedy plants and animals from surrounding agricultural lands and are vulnerable as well to perturbations created by agricultural production practices (Fry, 1995).

In peasant dominated areas, the use of traditional farming practices with minimal industrial inputs has resulted in a varied, highly heterogeneous landscape-possibly even more heterogeneous than would exist naturally. In such heterogeneous environments, natural and semi-natural ecosystem patches included in the landscape can become a resource for agroecosystems. An area of non-crop habitat adjacent to a crop field, for example, can harbor populations of natural enemies which can move into the field and parasitize or prey upon pest populations. (Altieri, 1994) A riparian corridor vegetated by native plant species can filter out dissolved fertilizer nutrients leaching from crop fields, promote a presence of beneficial species, and allow the movement of native animal species into and through the agricultural components of the landscape.

On the other hand, agroecosystems can begin to assume a positive rather than a negative role in preserving the integrity of natural ecosystems. Many small scale-diversified agroecosystems have been designed and managed in ways that make them more friendly to native species. For example, by encouraging hedgerows, vertebrates can be provided with large habitats, better food sources, and corridors for movement. Native plants can have more suitable habitats and find fewer barriers to dispersal. Smaller organisms, such as below ground microbes and insects, can flourish in organically managed soils and thus benefit other species since they are such important elements in ecosystem structure and function (Glissman, 1998).

By managing agricultural landscapes from the point of view of biodiversity conservation as well as sustainable production, the multiple use capacity of agriculture can be enhanced providing several benefits simultaneously (Thrupp, 1998):

- increase agricultural productivity;
- build stability, robustness, and sustainability of farming systems;
- contribute to sound pest and disease management;
- conserve soil and increase natural soil fertility and soil health
- diversify products and income opportunities from farms;
- add economic value and increase net returns to farms;
- reduce or spread risks to individuals, communities, and nations;

- increase efficiency of resource use and restore ecological health;
- reduce pressure of agriculture on fragile areas, forests, and endangered species;
- reduce dependency on external inputs, and;
- increase nutritional values and provide sources of medicines and vitamins.

The effects of agrobiodiversity in mitigating extreme climatic effects, such as the drought promoted by El Niño's were recently evident in northern Honduras. An agroforestry project reviving the Quezungal method, an ancient agricultural system, speared about 84 farming communities from destruction. Farmers using the method lost only 10 percent of their crops in 1998's severe drought, and actually obtained a grain surplus of 5-6 million pounds in the wake of Hurricane Mitch. On the other hand,, nearby communities which continued the use of slash and burn, were severely affected by El Niño phenomena, which left a legacy of human misery and destruction of vitally important watersheds.

Such agroforestry programs which reduce deforestation and burning of plant biomass can provide a sink for atmospheric carbon dioxide and also considerably reduce emissions of nitrous oxide. Recent research shows that promoting techniques already familiar to thousands of small farmers in Latin America such as, crop rotation and cutting back on chemical fertilizers through the use of composting can act as important sinks for atmospheric carbon dioxide storing it below the soil surface.

The benefits of agrobiodiversity in enhancing the multifunctional agriculture extend beyond the above described effects as shown by the impacts of shaded coffee farms in Latin America. Farmers typically integrate into their coffee farms many different leguminous trees, fruit trees, and types of fuel wood and fodder. These trees provide shade, a habitat for birds and animals that benefit the farming system. In Mexico, shade coffee plantations support up to 180 species of birds, including migrating species, some of which play key roles in pest control and seed dispersal.

Learning how to manage an agriculture that promotes both environmental as well as productive functions will require inputs from disciplines not previously exploited by scientists, including agroecology, ethnoscience, conservation biology, and landscape ecology. The bottom line, however, is that agriculture must adopt ecologically sound management practices, including diversified cropping systems, biological control and organic soil management as replacements for synthetic pesticides, fertilizers, and other chemicals. Only with such foundation can we attain the goal of a multifunctional agriculture.

Biodiversity and Pest Management

Nowhere are the consequences of biodiversity reduction more evident than in the realm of agricultural pest management. The instability of agroecosystems becomes manifest as the worsening of most insect pest problems is increasingly linked to the expansion of crop monocultures at the expense of the natural vegetation, thereby decreasing local habitat diversity (Altieri and Letourneau, 1982). Plant communities that are modified to meet the special needs of humans become subject to heavy pest damage and generally the more intensely such communities are modified, the more abundant and serious the pests. The effects of the reduction of plant diversity on outbreaks of herbivore pests and microbial pathogens is well-documented in the agricultural literature (Andow, 1991; Altieri, 1994). Such drastic reduction in plant biodiversity and the resulting epidemic effects can adversely affect ecosystem function with further consequences on agricultural productivity and sustainability.

In modern ecosystems, the experimental evidence suggests that biodiversity can be used for improved pest management (Altieri and Letourneau, 1994: Andow, 1991). Several studies have shown that it is possible to stabilize the insect communities of agroecosystems by designing and constructing vegetational architectures that support populations of natural enemies or that have direct deterrent effects on pest herbivores. For example, at the landscape level, data demonstrates that there is enhancement of natural enemies and more effective biological control where wild vegetation remains at field edges and in association with crops (Altieri, 1994). These habitats may be more important as overwintering sites for predators or they may provide increased resources such as pollen and nectar for parasitoids and predators form flowering plants (Landis, 1994). Many studies have documented the movement of beneficial arthropods from margins into crops and higher biological control is usually observed in crop fields close to wild vegetation edges than in fields isolated from such habitats (Altieri, 1994).

In many cases, weeds and other natural vegetation around crop fields harbor alternate hosts/prey for natural enemies, thus providing seasonal resources to bridge gaps in the life cycles of entomophagous insects and crop pests (Altieri and Letourneau, 1984). A classic case is that of the egg parasitoid wasp *Anagrus epos* whose effectiveness in regulating the grape leafhopper *Erythroneura elegantula* was increased greatly in vineyards near areas invaded by wild blackberry (*Rubus* sp.). This plant supports an alternative leafhopper (*Dikrella cruentata*) which breeds in its leaves in winter. Recent studies show that French prune orchards adjacent to vineyards provide overwintering refuges for *Anagrus* and early benefits of parasitism are promoted in vineyards with prune trees plants upwind from the vineyard.

At the crop field level, most experiments that have mixed other plant species with the primary host of a specialized herbivore show that in comparison with diversified cropping systems, monocultures have greater population densities of specialist herbivores (Andow, 1991). In these monoculture systems, herbivores exhibit greater colonization rates, greater reproduction, higher tenure time, less disruption of host finding and lower mortality by natural enemies.

There are various factors in crop mixtures that help constrain pest attack. A host plant may be protected from insect pests by the physical presence of other plants that may provide a camouflage or a physical barrier. Mixtures of cabbage and tomato reduce colonization by the diamond-back moth, while mixtures of maize, beans, and squash have the same effect on chrysomelid beetles. The odors of some plants can also disrupt the searching behavior of pests. Grass borders repel leafhoppers from beans and the chemical stimuli from onions prevent carrot fly from finding carrots (Altieri, 1994).

Alternatively, one crop in the mixture may act as a trap or decoy- the 'fly-paper-effect'. Strips of alfalfa interspersed in cotton fields in California attract and trap Lygus bugs. There is a loss of alfalfa yield, but this represents less than the cost of alternative control methods for the cotton. Similarly, crucifers interplanted with beans, grass, clover, or spinach are damaged less by cabbage maggot and cabbage aphid. Another factor as predicted by the natural enemies hypothesis is that reduced insect pest incidence in polycultures may be the result of increased predator and parasitoid abundance and efficiency (Altieri, 1994).

Most research conducted on traditional and peasant agriculture in Latin America suggests that small holder systems are sustainably productive, biologically regenerative, and energy-efficient, and also tend to be equity enhancing, participatory, and socially just. Besides crop diversity, peasant farmers use a set of practices that cause minimal land degradation. These include the use of terraces and hedgerows in sloping areas, minimal tillage, small field sizes, and long fallow cycles. By concentrating on short rotations and fewer varieties, agricultural modernization in the same areas has caused environmental perturbation and eroded genetic diversity (Altieri 1991, Altieri 1996, Wilken 1997).

By adopting a multiple use strategy, indigenous farmers manage a continuum of agricultural and natural systems, obtaining a variety of products as well as ecological services thus truly enacting a multifunctional agriculture. Diversified cropping systems, such as those used by peasants, based on intercropping and agroforestry have been the target of much research recently. This interest is largely based on the new emerging evidence that these systems are more sustainable and more resource- conserving (Vandermeer, 1995). These attributes are connected to the higher levels of functional biodiversity associated with complex farming systems. In fact, an increasing amount of data reported in the literature, documents the effects that plant biodiversity has on the stabilization of agroecosystem processes.

In a recently conducted, well replicated experiment, where species diversity was directly controlled in grassland systems, it was found that ecosystem productivity was increased and that soil nutrients were utilized more completely when there was a greater diversity of species, leading to lower leaching losses from the ecosystem

(Tilman et al. 1996). In traditional agroecosystems this same pattern applies to insects as herbivore regulation increases with increasing plant species richness. Evidence suggests that as plant diversity increases, pest damage seems to reach acceptable levels, thus resulting in more stable crop yields. Apparently, the more diverse the agroecosystem and the longer this diversity remains undisturbed, the more internal links develop to promote greater insect stability. One aspect that is clear is that species composition is more important than species number per se. The challenge is to identify the correct assemblages of species that will provide through their biological synergisms key ecological services such as nutrient cycling, biological pest control, and water and soil conservation (Altieri 1994).

While it may be argued that peasant agriculture generally lacks the potential of producing meaningful marketable surplus, it does ensure food security. Many scientists wrongly believe that traditional systems do not produce more because hand tools and draft animals put a ceiling on productivity. Productivity may be low but the causes appear to be more social, not technical. When the subsistence farmer succeeds in providing food, there is no pressure to innovate or to enhance yields. Nevertheless, NGO-led agroecological field projects show that traditional crop and animal combinations can often be adapted to increase productivity when the biological structuring of the farm is improved and labor and local resources are efficiently used (Altieri 1995). In fact, most agroecological technologies promoted by NGOs can improve traditional agricultural yields increasing output per area of marginal land, enhancing also the general agrobiodiversity and its associated positive effects on food security and environmental integrity (Pretty 1997).

It is not a matter of romanticizing subsistence agriculture or considering development per se as detrimental. The idea is to stress the value of traditional agriculture in the preservation of native crop diversity and the adjacent vegetation communities as this mode of appropriation of nature enhances the multifunctionality of agriculture (Toledo, 1995). Basing a rural development strategy on traditional farming and ethnobotanical knowledge, combined with elements of modern agroecology, not only assures continual use and maintenance of valuable agrobiodiversity, but also allows for the diversification of agricultural areas ensuring a variety of ecological services vital for food security, natural resource conservation, economic viability, climate amelioration, cultural preservation, and community empowerment. The challenge is now to promote the right policies and institutional partnerships that can scale-up ecologically based agriculture so that its multifunctional impacts are rapidly spread across the rural landscapes of Latin America.

Ecological Impact of Modern Agriculture

We might call this the Age of the Recognition of Limits. A generation ago, the energy reserves that fueled the extractive economy were seen to be the primary limits affecting human progress in every endeavor, including agriculture. American scientists and policymakers alike were comfortable in their belief that these acknowledged finite reserves were large enough to buy the time necessary to bring the peaceful atom on line. It isn't that renewable resources were ignored. Americans did worry about the consequences of soil erosion and water pollution, of course, but there was a nearly full confidence in the technical fix for problems in the renewable sector of the economy. For agriculture, this meant that terraces, grass waterways, and proper incentives would take care of erosion. Simple solutions like indoor plumbing and chlorine in our drinking water would handle the water problem. And as for this nonsense about pesticides, Rachel Carson was just a hysterical extremist who was not an expert. The attitude about commercial fertilizers was that "nitrogen is nitrogen" and "plants don't care whether it comes from an anhydrous ammonia tank or from the nodule of a legume." And besides that, "We must feed the world." We had a sense of the heroic urge in us. It was also an era in which "hard-headed realism" was a favorite expression. But as author Wendell Berry has said, "A hard-headed realist is usually a person who uses a lot less information than is available." We have more information now than a generation ago, and the public is better informed.

The "problem of agriculture" is as old as agriculture itself, and although the core of the problem has always been soil erosion, new problems have been added. The high-energy epoch that fueled industry, which in turn made the industrialization of agriculture possible, simply exacerbated the old problems in agriculture and added more besides. And so we are dealing with an old subject with new data and the need for a fresh approach in our search for solutions. We will consider these problems one at a time even though they are inextricably intertwined.

Soil Erosion

Raindrops bombarding bare soil result in the oldest and still most serious problem of agriculture. The long history of soil erosion and its impact on civilization is one of devastation. Eroded fields record our failure as land stewards. W. C. Lowdermilk (1953) described several civilizations that collapsed because of erosion. At the end of his historical sweep of failed civilizations, Lowdermilk warned that unless we want to experience a similar fate, we had better heed these lessons of the past and safeguard our soils.

Soil erosion came to the nation's attention dramatically during the dust-bowl 1930s. Dust storms that blackened the sky were harder to ignore than the more severe problem of sheet and rill erosion due to rain. The Soil Conservation Service (SCS) was formed during this period under the energetic and imaginative direction of Hugh Hammond Bennett. The SCS encouraged planting shelterbelts of trees and grassed waterways and using contour ploughing and terracing. But by the mid-1970s soil erosion once again became a national concern. Farmers were encouraged both by the export potential and by Secretary of Agriculture Earl Butz to plant from fencerow to fencerow (Weasel, 1983), bringing 24 million additional hectares (60 million acres) into production (Crosson and Stout, 1983; Batie, 1983). By and large, the extra land was more prone to erosion. Larger machinery was purchased and pulled into the fields, and suddenly the terraces on sloping land and the windbreaks were nuisances (Weasel, 1983). Soil conservation meant increased costs to farmers, who were already burdened by the additional investments. The possibility of higher profits overshadowed any long-term benefits of conservation.

In 1977 the SCS, in an attempt to understand the amount of this precious resource being lost, began a new survey called the National Resources Inventory (NRI), published in 1981. The public was given a comprehensive review of sheet, rill, and wind erosion in the United States over a cropland base of 167 million hectares (413 million acres) (Crosson and Stout, 1983; Batie, 1983). An estimated total of 5.8 billion metric tons (6.4 billion tons) of topsoil was washed or blown away in the United States (USDA, 1980; Crosson and Stout, 1983). Some 2 billion metric tons of soil was lost from American cropland alone, an average of 15.4 metric tons/ha (6.8 tons/acre) (Crosson and Stout, 1983). Sheet and rill erosion were responsible for 1.7 billion metric tons being lost, while wind swept away 816 million metric tons (Crosson and Stout, 1983). The average “official” tolerable losses of 11.3 metric tons/ha (5 tons/acre) for deep soils and of 2.3 metric tons/ha (1 ton/acre) for shallow soils were violated (Crosson and Stout 1983; Batie, 1983).

Since erosion is not evenly distributed, in some areas the erosion rate greatly exceeded 15.4 metric tons/ha (6.8 tons/acre) In the Palouse Hills region of southeastern Washington and southwestern Idaho, where dryland farming is devoted to wheat, barley, beans, and lentils, there are numerous slopes ranging in steepness from 15 to 25 percent; here the soil erosion due to snow and rain was 113 to 227 metric tons/ha (50 to 100 tons/acre) (USDA, 1980). In southeastern Idaho, where hard red wheat is planted on slopes as steep as 35 percent, the annual erosion rate was 36 metric tons/ha (16 tons/acre) (Batie, 1983). Steep but fertile ground in southern Mississippi experienced a loss of 45 metric tons/ha (20 tons/acre) or more (USDA, 1980). Corn-belt losses up to 45 metric tons/ha were common (Batie, 1983).

Although the NRI was the most extensive survey yet produced, it did not give a complete picture of the amount of soil loss. The NRI relied on the universal soil loss equation (USLE) and the wind erosion equation (WEE) for estimating total soil loss (Batie, 1983). The USLE can estimate sheet and rill erosion, but it will not determine the amount of soil lost in gullies (Crosson and Stout, 1983). Wind erosion was determined for only 10 states in the 1977 survey (Crosson and Stout)

The more comprehensive NRI in 1982 documented an increase in the nat~on's cropland base from 167 million to 170 million hectares (413 million to 421 million acres) during the 5 years between surveys (Lee, 1984). The acreage in the land-capability classes and subclasses most subject to erosion increased disproportionately. The size of the highly erodable land-capability classes of V, VI, VII, and VIII increased 9 percent, while cropland in the high-quality classes I, II, and III increased only 1.2 percent (Lee, 1984). The sum of wind and water erosion averaged more than 18 metric tons/ha (8 tons/acre), or 3.1 billion metric tons (3.4 billion tons) on cropland. Forty-four percent of the cropland lost soil above the tolerance level (Lee, 1984). Nearly a fifth of the acreage was devoted to corn, a crop linked to high erosion rates (Bills and Heimlich, 1984).

Global losses are also high. Lester Brown (1981) reports that around 21 billion metric tons of soil is being lost worldwide. Even India and China, presently self-sufficient in food production, are losing soil beyond replacement levels. In India an estimated 6 billion metric tons of soil is lost annually from their croplands, representing a 4.7 billion metric tons/yr loss beyond the tolerance level (Brown, 1984).

Erosion is not always obvious, especially sheet erosion. Three steps characterize the erosion process: Soil particles are detached, then displaced by wind and rain, and finally deposited. Raindrops striking the bare soil can throw soil particles several centimeters. Sheet erosion, or overland flow, occurs when the water removes a relatively uniform thickness of soil. Rill erosion is easier to detect, since we can readily see the small channels, or rills, where water concentrates. The erosive power of the rill variety is responsible for a majority of our soil losses. (All from Morgan, 1979.)

Wind erosion also separates soil particles. Fine particles less than 0.2 mm (0.01 in) in diameter are transported high in the air over long distances (Lyres and Tatarko, 1986). Larger particles roll along the ground, become abraded, and add to the fine particles that are removed by the wind (Lyres and Tatarko, 1986; Morgan, 1979). Soil structure is damaged and soil texture altered as the coarser particles are left behind and organic matter decreases (Lyres and Tartako, 1986).

With such a wholesale altering of its basic structure, texture, and organic matter, the deposited soil no longer has the potential to produce its former yields (Rosenberry et al., 1980; Meilke and Schepers, 1986; National Soil Erosion-Soil Productivity Research Planning Committee, 1981). As the soil erosion continues, the subsoil becomes part of the tillage layer, and the fertilizer inputs and power requirements increase to the point where it is no longer feasible to cultivate the land (Rosenberry et al., 1983).

The problems of soil erosion are not limited to the farm. Soil from eroded cropland is the largest contributor to nonpoint pollution in the United States (Meyers et al., 1985). Sediment makes up the greatest volume by weight of materials transported in streams and rivers, and other pollutants (such as pesticides and fertilizers) can be carried along with it, either adsorbed on the soil particles or in solution (Clark, 1985).

These pollutants adversely affect aquatic ecosystems either by directly affecting fish and other aquatic life or by altering spawning areas and food sources (Clark, 1985; Meyers et al., 1985). In addition, sediment accumulation is reducing the capacity of waterstorage facilities; approximately 1727 m3 (1.4 million acre-feet) of reservoir and lake capacity is filled each year with sediment (Clark, 1985). Dredging is a costly way of maintaining this capacity, and alternative reservoir sites are becoming difficult to find (Poster, 1985). The consequence is that municipal water sources are affected. The costs to clean the water are substantially higher since cleaning devices (sedimentation basins, filters, and chemical coagulants) need to be installed (Clark, 1985).

The USDA has promoted no-till and minimum-tillage practices as erosion control measures. In 1982, 4.7 million hectares (11.6 million acres) of farmland were in no-till, and the USDA has predicted that 85 percent of all cultivated cropland will be in some form of minimum tillage by the year 2000 (Hinkle, 1983). Unfortunately, no-till practices rely heavily on herbicides.

Some promising attempts to reduce the amount of soil erosion are the two conservation provisions of the 1985 Farm Bill (Wenzel, 1986). Under the first, the SCS will identify highly erodable land. Farmers who convert any highly erodable land to cropland will be denied such benefits as price supports and crop insurance (Wenzel, 1986), possibly keeping 13 million hectares (32 million acres) from being converted to cropland. The second provision, the Conservation Reserve, pays farmers to take highly erodable land out of production; it caused 3.6 million hectares (9 million acres) to be taken out of production in 1986 (American Farmland Trust, 1986). By 1986, Congress had a target of 18.2 million hectares (45 million acres) (American Farmland Trust, 1986), but the number is likely to go higher.

Future efforts at soil conservation should concentrate on devploping farming practices that preserve and enhance the soil. Farmers should be subsidized to initiate soil-conserving practices, such as terracing and contour plowing. Also, we should look to natural ecosystems to provide our best example, such that soil surfaces are always protected by a canopy. This is difficult to obtain in an annual cropping system where the soil is tilled every year. Reduced-tillage practices have helped conserve soil, but the use of herbicides with this practice makes it undesirable in a sustainable agriculture. Perhaps in our search for new and more sustainable agricultural methods we will find other ways to conserve soil. Planting perennial cropping systems on sloping ground may allow continued use of this fragile land, or we may have to convert it to grassland and limit our cropping systems to flat bottomlands.

Irrigation

Adequate rainfall is never guaranteed for the dryland farmer in arid and semiarid regions, and thus irrigation is essential for reliable pro auction. Irrigation ensures sufficient water when needed and also allows farmers to expand their acreage of suitable cropland. In fact, we rely heavily on crops from irrigated lands, with fully one-third of the world's harvest coming from that 17 percent of cropland that is under irrigation (Poster, 1985). Unfortunately, current irrigation practices severely damage the cropland and the aquatic systems from which the water is withdrawn and partially returns.

Currently, the 255 million irrigated hectares (631 million acres) in the world use 70 percent of the total world water consumption (Poster, 1985). In the United States more than 20 million irrigated hectares (49 million acres), concentrated in the 18 western states and in the southeast, use 83 percent of the total water consumed (Fredrick and Hanson, 1982; USDA, 1985). Very few conservation efforts are currently devoted to water, probably because it has been regarded as a limitless resource. Farmers worldwide face the problem of limited supplies and the resultant competition over what is left. But even in the face of decreased supplies, the number of irrigated acres continues to expand. In the 1970s an additional 5.1 million hectares (12.5 million acres) per year were irrigated (Poster, 1985). Water from underground aquifers supplies 40 percent of our irrigation needs in the United States (Pimentel et al., 1982). Groundwater provides an excellent high-quality source. In the United States the storage capacity of aquifers greatly exceeds that of rivers, lakes, and reservoirs (U.S. Geological Survey, 1984). But supplies are not limitless; aquifers recharge slowly from water percolating through the soil, and water in some deep aquifers can be considered a nonrenewable resource. The amount of water that is pumped from the aquifers exceeds the amount replenished in many areas of the world. In the United States approximately one-fourth of the groundwater extracted exceeds the replenishment (Stokes, 1983). As water tables fall, springs dry up and streams experience reduced flow (Poster, 1985; Pimentel et al., 1982).

The Ogallala aquifer is a good example of rapid aquifer depletion. This aquifer lies under parts of seven states: Texas, New Mexico, Colorado, Kansas, Nebraska, Wyoming, and South Dakota. It supports one-fifth of the U.S. irrigated cropland (Poster, 1985). The export push in the 1970s encouraged farmers to install extensive irrigation systems (Weasel, 1983). Farmers irrigated 3.2 million hectares (8 million acres) in 1978, compared to 0.85 million hectares (2.1 million acres) in the 1940s (Poster, 1985). A center-point irrigation system (the type now prevalent in this area) can use 3030 L (800 gal) of water per minute (Adams, 1981). Five hundred cubic kilometers (405.4 million acre-feet) of water has been withdrawn in the last four decades, and some hydrologists predict that given the current rate of drawndown, the aquifer will fail to yield in another 40 years (Ferguson, 1983). The exhaustion of the Ogallala aquifer is showing its effects; land in Texas, New Mexico, and Kansas is being taken out of production because of diminishing well yields and rising pumping costs (Ferguson, 1983). A total of 6 million hectares (15 million acres) in 11 states may be taken out of production because of water shortages; Texas alone may lose 1.2 million hectares (3 million acres) by the year 2000 (Stokes, 1983).

Declining water levels not only affect farmers. In the southeast, several cities depend on groundwater to meet their water needs. Tucson, Arizona, is a city completely dependent on an aquifer to meet its needs, yet only 35 percent of the water withdrawn is replaced by recharge (Stokes, 1983). Exhausted water resources will effectively curtail the expansion of these cities.

Land subsidence, the dropping of the land surface, is another major problem when aquifers are overdrawn (Bouwer, 1981). Numerous incidences of land subsidence have been recorded in California (Carbognin, 1985). Heavy pumping of the groundwater began after World War II and has continued to the present (Carbognin, 1985). In the 51 years between 1925 and 1977, 20 billion cubic meters (26.2 billion cubic yards) of land has dropped due to the extraction of 70 million cubic meters 57,000 acre-feet of groundwater (Carbognin, 1985). Groundwater withdrawal in Mexico City has caused 9 m (30 ft) of land subsidence, creating problems with their drainage system and with construction (Carbognin, 1985). In addition to subsidence, the depletion of aquifers along the coastlines of California, Georgia, and Florida has caused saltwater intrusions (Pimentel et al., 1982).

Depletion of streams and rivers is also a serious problem. River ecosystems require minimum levels of instream flow. Aquatic and riparian habitats are severely affected or destroyed when rivers are overdrawn (U.S. Water Resources Council, 1978). In the l8 western states most dependent on irrigation, 70 percent of the instream flow is often depleted (Fredrick and Hanson, 1982). For example, in the lower Colorado River more than 80 percent of the instream flow is depleted (Pimentel et al., 1982). River runoff may not be adequate for optimal fish and wildlife habitats in rivers of 14 western and southwestern states by the end of this century (Stokes, 1983). To protect these natural habitats, irrigation would have to be reduced to one-fifth of what it is now (Stokes, 1983).

Beyond depleting water resources, current irrigation systems can destroy cropland and reduce water quality. Waterlogging, salinization, and alkalinization have damaged about half of the world's irrigated lands (CEQ, 1980). Although this is only a small percentage of the world's total cropland, the losses are significant when the higher yields obtained with irrigation are considered. The soil and groundwater in arid zones typically have a high salt content, and natural drainage is usually poor (Kovda, 1980). Irrigation water drains slowly from fields, and as the water table rises, the soils become waterlogged in the crop root zone (Kovda, 1980). Continued application of irrigation water and seepage from unlined canals cause the salty groundwater to reach the surface (Kovda, 1980). As the water evaporates from the soil, a salty white crust forms on the surface, leaving the land unfit for cultivation unless expensive reclamation procedures are used. These problems are present wherever large acreages of cropland are irrigated (Kovda, 1977). Pakistan has enormous problems with salinization; 9.1 million of its 14.6 million irrigated hectares (22.4 million of 36 million acres) have some form of salinity problem (CEQ, 1980). In the San Joaquin valley about 162 million hectares (400,000 acres) are affected by high brackish water tables. Unless adequate drainage is installed, approximately 13 percent of this acreage will become unproductive (Poster, 1985).

Cropland salinity also creates water-quality problems in arid and semiarid river basins. Less than 50 percent of the water applied for irrigation is used by plants; the rest eventually returns to rivers and streams (Poster, 1985). The quality of the return flow is seriously degraded by salts, sediments, fertilizers, and pesticides. In addition, groundwater sources can become contaminated by the deep percolation of degraded irrigation water (El-Ashry et al., 1985). As a result, municipal and industrial water sources become polluted and users must pay higher water-treatment costs.

Applying large amounts of water in arid regions can also alter the ecosystem; in addition, such changes in farming practices as plowing, fertilizing, and monocropping (which usually accompany irrigation) also affect the ecosystem. Plant diseases, insects, and weeds, previously unable to withstand the dry conditions, can become serious pests (Ghabbour, 1977). The soil fauna, usually sparse in desert climates and adapted to drought conditions, can undergo drastic changes with irrigation; certain groups of soil fauna become dominant, destroying the previous balance. For example, earthworms, collemba, mites, and nematodes will invade newly irrigated soils at the expense of the original fauna unable to adapt to the new moisture levels (Ghabbour, 1977).

Human diseases have also increased with the widespread use of irrigation, especially water-transmissible diseases such as malaria and schistosomiasis (Obeng, 1977; Farid, 1977). The aquatic snail that transmits schistosomiasis inhabits sluggish waters common in irrigation projects, and schistosomiasis is now estimated to affect 200 million people in Asia, Africa, the Caribbean, and South America (Obeng, 1977). The disease has reached its current levels with the construction of irrigation projects. In the Gezira project in the Sudan, for example, the incidence of the disease increased almost 50 percent in a 20-year period after the expansion of the irrigation project in 1950 (Obeng, 1977).

Although irrigation in arid areas cannot be totally abandoned, we need to realize that water is a limited resource and that the arid ecosystems are fragile. Upgrading irrigation systems will ameliorate some of the problems outlined above. Instead of financing massive water-diversion projects, improving irrigation efficiencies can free water for other demands. Lining canals, reusing water runoff, and increasing the efficiency of flood and furrow systems can save enormous quantities of water. Unfortunately, the incentive to save water is lacking in the United States since farmers pay only one-fifth of the cost of the water they use (Ferguson, 1983). Limited supplies and rational costs will force changes in the allocation and use of water.

In addition, the type and extent of agriculture practiced in arid regions should be reexamined. Agricultural systems transferred from humid areas are not suitable. Investigating the unique qualities of arid ecosystems will help develop more appropriate agricultural practices.

Agriculture and the Loss of Genetic Diversity

As modern agriculture converts an ever-increasing portion of the earth's land surface to monoculture, the genetic and ecological diversity of the planet erodes. Both the conversion of diverse natural ecosystems to new agricultural lands and the narrowing of the genetic diversity of crops contribute to this erosion.

An estimated 852 million hectares (2.1 billion acres) worldwide have been converted to regular cropping or grazing in the past 120 years through deforestation, draining wetlands (or converting to rice), irrigation, and conversion to grazing (Richards, 1984; Salick and Merrick, Chap. 19 of this book). Although in North America from 1920 to 1978 the reversion of cultivated land to wild land about equaled the conversion to new agricultural land, elsewhere conversion has far exceeded reversion. In Africa 90.5 million hectares (223.6 million acres) were converted during this same period, and worldwide net conversion (conversion minus reversion) equaled about 420 million hectares (1.04 billion acres) (Richards, 1984). Current world population trends and the unequal distribution of power and resources will keep the pressure for continued conversion high (Wilson, 1985). In the United States the increased demand for exported food in the 1970s caused the conversion rate to climb dramatically; an estimated 24 million hectares (59 million acres) were converted to agriculture between 1972 and 1980 (Richards, 1984). Inevitably, some ecologically sensitive areas, such as the vital breeding grounds for wetland birds and waterfowl known as the "prairie potholes" in North Dakota, were victims of this agricultural expansion.

In the tropics, rates of land conversion to agriculture remain high (Richards, 1984), and there the impact on global species diversity is the greatest. The ongoing destruction of tropical forests is estimated to cause thousands of species extinctions annually (Wilson, 1985). Scientists estimate that 3 million species of plants (two-thirds of all species on earth) reside in the tropics, and about half of these are confined to tropical forests (U.S. Department of State, 1981; White, 1983). At the current rates of deforestation, one-fourth to one-third of all modern species will be gone in the next 30 to 40 years (U.S. Department of State, 1981; Batten, 1983).

The pressure on tropical forests is twofold. On the one hand, the need for foreign exchange, coupled with the industrial nations' voracious consumption of forest products, produces lucrative economic incentives for logging operations (World Commission on Environment and Development, 1987; Nations and Komer, 1983). Historically, economic factors have been the prime cause of most deforestation (Richards, 1984). On the other hand, inequitable land distribution has created a rapidly growing population of landless peasants. In Latin America a mere 7 percent of the landowners control 93 percent of the arable land. From Brazil and Colombia to Kenya, Thailand, Indonesia, and Madagascar the story is similar. Logging roads provide access to unclaimed lands, and once the valuable timber species have been logged out, the landless poor are encouraged to move in and clear the land. In the process, indigenous peoples are likely to be displaced, thus increasing the pool of landless poor. Typically, the cleared land will yield for only 5 to 7 years of intensive cropping, and then the peasants must look for a new plot. Often the newly cleared land is taken over by large landowners to raise export crops, such as beef in Latin America, at pitifully low productivity rates (World Commission on Environment and Development, 1987; Nations and Komer, 1983). Grazing cattle in cleared rain forest in Chiapas, Mexico, yields a mere 10 kg/ha (8.9 lb/acre) of meat per year. In sharp contrast, the traditional methods of the indigenous Maya yield 5896 kg/ha (5263 lb/acre) of shelled corn plus 4535 kg/ha (4049 lb/acre) of roots and vegetables per year for 5 to 7 years, then additional yields of citrus, rubber, cacao, avocado, and papaya for the following 5 to 10 years while the plots are allowed to regrow (Iltis, 1983). To remove the pressure on tropical forests, a complex set of changes must occur, ranging from (1) economic disincentives to the industrial nations' importation of tropical wood to (2) extensive agrarian reforms in the developing nations. These reforms include land redistribution that returns productive land to displaced peasants, the development and adoption of ecologically sound farming practices, and the goal of self-sufficiency in order to reduce the need for foreign exchange.

Modern agriculture's impact on the genetic diversity of crop plants is also profound. Recent gains in yield have been achieved by increased dependence on very few genetically narrow cultivars of only a handful of major food plants (Committee on Germplasm Resources, 1978; N. J. Brown, 1981; Mayer, 1981; Wilson, 1983; Duvick, 1984; Kannenberg, 1984). Only 10 to 20 crops provide 80 to 90 percent of the world's calories (N. J. Brown, 1981; Mayer, 1981)—or as W. L. Brown (1981) states, 15 species of cultivated plants "literally stand between man and starvation." As of 1970, in the United States 56 percent of the soybean crop, 71 percent of the hybrid. corn crop, and 41 percent of wheat acreage were occupied by only six cultivars each (Duvick, 1984). For red winter wheat, only two varieties made up 75 percent of the wheat land, with a single variety covering more than 50 percent of the acreage. Although the dominance of the top six cultivars was reduced somewhat by 1980—to 42 percent of soybeans, 43 percent of corn, and 38 percent of wheat (Duvick, 1984)—still, the genetic base of the varieties is limited (Kannenberg, 1984). The major cultivars of red winter wheat trace back to only two original cultivars: Turkey and Marquis. The hundreds of corn hybrids grown in the United States and Canada are largely based on about 12 inbred lines that originated from a few open-pollinated varieties of a single race (out of some 200 known races of corn). The U.S. soybean varieties were derived originally from six plants from Asia (N. J. Brown, 1981).

Throughout the world the attraction of high yields from new strains has caused farmers to abandon the more diverse, locally adapted varieties, or landraces (Committee on Germplasm Resources, 1978; Pluckiest et al., 1983; Van Sloten, 1984). For example, F1 hybrids have virtually replaced landraces of cultivated Brassica spp. in Japan, Korea, North and South America, and Northern Europe. China, the last holdout for genetic diversity of this group, is now losing landraces rapidly (Van Sloten, 1984). In the past 40 years, 95 percent of Greek wheat landraces have been lost because of the introduction of highyielding varieties (Pluckiest et al., 1983). Similarly, nearly all sorghum landraces disappeared from South Africa following the introduction of high-yielding hybrids fróm Texas (Pluckiest et al., 1983). Landraces of white potatoes, sugarcane, and tomatoes are already mostly lost (Committee on Germplasm Resources, 1978). The problem is compounded by governmental policy where official seed agencies emphasize uniformity and purity of seeds, as in the United States and Canada (Committee on Genetic Vulnerability of Major Crops, 1972). Solutions to this threat will need to be multifaceted.

Reduced genetic diversity in crops results in a loss of flexibility to meet future breeding challenges. Ironically, the landraces in jeopardy are essential to the maintenance of the high-yielding, genetically narrow cultivars that replace them (Committee on Germplasm Resources, 1978; Bertrand, 1981; Pluckiest et al., 1983; Smartt, 1984). Landraces are the breeders' sources of resistance to diseases and pests and of adaptation to climatic extremes and poor soils (Brady, 1981; Pluckiest et al., 1983; Kannenberg, 1984). For example, the source of powdery mildew resistance in California melons was a wild melon from India. Resistance to ripe rot in North American pepper plants came from a Peruvian species. High-yielding hybrid cucumber seed was obtained from a Korean strain (Bertrand, 1981). Wild genotypes of rice have helped to boost rice yields by an estimated $1.5 billion per year (Myers, 1983).

Several recent trends further increase the need for new sources of genetic diversity: Wide use of uniform varieties in monoculture has increased vulnerability to biological and physical stresses; expansion of agriculture into marginal lands requires special adaptation to climatic extremes, poor soil, and low water quality; consumer demands for appearance, flavor, and quality require continual innovation, as do such changes in farm technology as no-till farming, multiple cropping, and biological farming (Pimentel, 1977; Bertrand, 1981). Without the landraces and wild species to draw on in the future, breeders will reach the limits set by the variation available in the cultivars and have nowhere to turn for future innovation (Kannenberg, 1984).

Crop breeders have begun to respond to this dangerous loss of genetic diversity. Reduction in dependence on the top six cultivars of wheat, soybeans, and corn between 1970 and 1980 in the United States is one example (Duvick, 1984). Corn breeding has shifted most dramatically toward genetically broader varieties (Kannenberg, 1984). Pedigree breeding (where the best cultivars of each generation are the parents for the next generation), which results in a very narrow genetic base, is less popular now than it once was. By 1975 synthetic populations (intercrosses of several elite inbred lines), crosses of exotic and adapted lines, and population-based recurrent selection programs were producing 49 percent of the new lines of corn. Kannenberg (1984) advocates adoption of an even richer genetic foundation, which he achieves by a hierarchical open-ended system in which new genotypes are continually introduced into the breeding program.

Widespread recognition of the severity of the genetic erosion of crop species came in the 1960s, and by the early 1970s this recognition had turned to alarm among crop breeders (Williams, 1984). The Consultative Group on International Agricultural Research (CGIAR) was formed in 1971 to study the problem, and in 1973 the CGIAR created the International Board for Plant Genetic Resources (IBPGR). This board undertook the major tasks of collecting, organizing, and preserving seed sources representing as much of the worldwide diversity of crop and related species as possible. The funding for this program is international, and the seeds and information gathered are meant to be freely available to all. Although this program achieved a great deal in its first 10 years, especially in the collection of lines of the major grain crops, in 1984 the job was still far from complete (Committee on Germplasm Resources, 1978; Williams, 1984). Systematic ecogeographic searches in remote areas are still needed, with special attention to disease and pest resistance. Closely related wild species and most tropical crops have been largely neglected. Effective methods for preserving vegetatively propagated crops and those with short-lived seeds have yet to be found. Even though many seed sources are now safely preserved, most are still essentially unavailable to breeders because they have not been evaluated or characterized. This huge job will require major effort and funding. In many cases, in situ conservation of wild crop relatives and landraces is necessary to effectively preserve their genetic diversity, particularly for perennial vegetatively propagated crops (Ingram and Williams, 1984). The most direct way to accomplish the in situ preservation of genetic diversity may be to preserve indigenous subsistence agriculture. Where local varieties are found to be nutritionally superior, ecologically sound, and more reliable in unpredictable growing conditions, these systems should be maintained rather than replaced with modern, genetically limited varieties and farming techniques..

Chemical Contamination

"In nearly all respects agriculture became an industry, sharing with the traditional manufacturing industries the problems of waste byproducts disposal." Young (1983) is referring in this quote to the changes in modern agriculture that have occurred over the past 45 years since the introduction of cheap inorganic nitrogen fertilizers. Six kilograms (13 lb) of pesticides, 63.5 kg (140 lb) of actual nitrogen, 19 kg (42 lb) of phosphate—these were the average chemical inputs per irrigable acre in California in 1980. In 1 year, California alone uses 55 million kilograms (121 million pounds) of restricted-use pesticides (Afford and Ferguson, 1982). Along with the promise of high production that these figures imply is the potential environmental pollution that these massive chemical inputs can cause.

Pollution of Groundwater and Surface Water

Nitrate is currently recognized as the most serious agricultural chemical pollutant (Alfoldi, 1983). As nitrate shows up in drinking water, where it can cause methemoglobinemia in infants, attention is increasingly focused on nitrates in the groundwater source. Agriculture contributes to groundwater nitrate through the leaching of nitrogen fertilizer, animal wastes from feed lots, and organic matter from plowing under grasslands and crop residues (Guenzi, 1974; Aldwell et al., 1983; Alfoldi, 1983; Young 1983). Studies in Illinois from 1969 to 1972 showed that 73 percent of farm wells in areas with shallow aquifers had nitrate levels above the critical level considered safe to drink (Singh and Sekhon, 1978/1979). Reports of elevated nitrate levels in groundwater from England, Israel, Germany, France, and the Netherlands surfaced in the 1970s, and at a recent symposium other European countries, as well as Nigeria and India, joined the growing list (Young, 1983).

Where water tables are shallow, nitrate can reach the water table quickly (Zaporozec, 1983). A number of studies have shown correlations between nitrogen fertilizer use and polluted groundwater (Singh and Sekhon, 1978/1979; Zaporozec, 1983; McWilliams, 1984). In addition, irrigation on porous, shallow soils exacerbates the problem of nitrate leaching. An experiment conducted in Minnesota on sandy glacial outwash soils above a 4.5-m (14.8-ft) water table clearly demonstrated an increase in nitrate in the groundwater after onetime applications of nitrogen fertilizer (Gerwing et al., 1979). Deeper water tables can also become contaminated by nitrate, but fewer cases are known. In a 1966 survey of thousands of rural Illinois wells, only 1.4 percent of the water samples from deep wells [deeper than 30 m (98 It)] exceeded the Environmental Protection Agency (EPA) limit of 10 mg/L (38 mg/gal), as compared to 23 percent of the samples from shallow wells. But in 1982 in rural Rock County, Wisconsin, 10 percent of the water samples from deep wells [deeper than 46 m (151 ft)] exceeded the limit (Zaporozec, 1983). In England, high nitrate concentrations were found to a depth of 35 m (115 ft) below the minimum water table (Young, 1983). The movement of nitrate into deep groundwaters occurs so slowly in relation to horizontal groundwater movement that it is difficult to prove a direct association with fertilizer applications (Singh and Sekhon, 1978/1979); even so, nitrate levels are usually correlated with overlying land use. One study found the highest nitrate concentrations below cattle feedlots and irrigated alfalfa, followed by fallow wheat and then by virgin grasslands, which had the lowest concentrations (Singh and Sekhon, 1978/1979). Decontamination will take place as slowly as contamination. Once below the root zone, nitrate is highly stable (Singh and Sekhon, 1978/1979) and can percolate slowly toward deep aquifers for years. Even if all fertilizer application were suspended immediately in areas where deep aquifers are beginning to show nitrate pollution, the inputs of past years would still slowly percolate toward the aquifers, and nitrate input to the groundwater would continue for many years.

Deep aquifers are fairly well protected from pesticide contamination (Alfoldi, 1983), since all but the most persistent pesticides degrade before they reach a deep aquifer. Shallow aquifers, however, are more commonly contaminated with pesticides. In the shallow aquifers of Wisconsin's Central Sands region, the pesticide aldicarb has reached the groundwater in amounts above the EPA limit (McWilliams, 1984). In California, agricultural use of DBCP, a highly leachable nematicide used on a variety of crops, contaminated wells until its use was suspended in 1977 (FAO, 1979; Alford and Ferguson, 1982; Lecos, 1984).

Surface waters receive agricultural chemicals by runoff from field applications and feedlots and from the dumping of excess chemicals (Afford and Ferguson, 1982). Phosphorus from fertilizer applications binds to soil particles, whereby it can be exported in erosive runoff. Soluble organic forms in animal wastes or plant residues can also move into surface waters. Phosphorus is often the limiting nutrient in aquatic systems and is thus a major contributor to eutrophication (Oglesby, 1971). Nitrate, which is highly soluble and easily transported in runoff, also contributes to eutrophication (Oglesby, 1971). In combination, these two fertilizers have contributed much to the degradation of streams and other bodies of water in agricultural regions.

Pesticide runoff is normally low, with losses of 0.5 percent or less (Wauchope, 1978), giving concentrations of 1 rLg/L (3.8 g/gal) or less (Guenzi, 1974). Under the worst conditions—a major rainfall or heavy irrigation within 3 days of application, the most mobile pesticides, sloping ground, and fine-textured soils—the losses may reach up to 15 percent; however, under these conditions the pesticide is likely to be in low concentration (Wauchope, 1978). Higher concentrations of pesticides in runoff occur when the runoff volume is low; even so, the concentrations are usually below the acute toxicity levels for aquatic organisms (Guenzi, 1974). Ironically, contaminated runoff is less likely to poison aquatic life than to damage other crops. Irrigation water contaminated with herbicides in one field can reduce crop productivity in another field. Picloram concentrations as low as 0.1 ppm can reduce sugar beet yield significantly (Guenzi, 1974).

The dangers of persistent pesticides in aquatic ecosystems, especially the organochlorines (e.g., DDT), have been known for many years. These chemicals, although not highly toxic at low concentrations, stay in the bodies of consumers and thus accumulate up the food chain (A. W. A. Brown, 1978; Guenzi, 1974). Top predators (such as fish-eating birds) consume toxic dosages with disastrous results. Agricultural use of toxaphene from 1957 to 1961 in the Klamath basin of northern California killed white pelicans and other fish-eating birds in wetland refuges (A. W. A. Brown, 1978).

As these consequences became known, and as many organochlorine compounds came to be banned in the richer countries, they began to be replaced with shortlived, highly toxic organophosphates (such as parathion). Although toxic in minute quantities, the short life of such chemicals makes them unlikely to accumulate in bodies of water, and it also makes them difficult to study (Alfoldi, 1983; Guenzi, 1974; Wauchope, 1978). So far, the presence of nonpersistent pesticides has been documented, but no evidence of permanent impact in aquatic systems has been found (Wauchope, 1978). Studies set up to determine the impact of these pesticides have been limited to the direct application of pesticides to streams at higher concentrations than those found in agricultural runoff. These studies show that aquatic organisms vary widely in their sensitivity to different pesticides. Studies of organophosphates have shown that stone flies are sensitive to malathion, but less so to trichlorfon. In one study, malathion caused only a temporary depression of stream insect fauna, whereas fenitrothion caused extensive kills of stream insects (A. W. A. Brown, 1978). Species-specific responses to particular chemicals have been used to infer possible damage from natural agricultural runoff. A stream community in North Carolina that was affected by agricultural runoff showed a pattern of species reduction that matched the pattern caused by the direct application of carbaryl to streams in other studies—a correspondence that led Lenat (1984) to infer that this pesticide may have been present in the runoff. So far, studies of the effects of long-term, low-dose exposure of aquatic systems to nonpersistent pesticides are rare to nonexistent. Such studies should be pursued in order to ensure that we avoid serious environmental consequences.

Residues on Food

Pesticides also pose hazards by direct contamination of foodstuffs. To what extent is our food contaminated with pesticide residues, and how much of a hazard is this?

The long-lived pesticides, such as DDT, other organochlorines, and parathion, are most likely to leave persistent residues (Bull, 1982). Even though most persistent pesticides are banned in the United States and other wealthy countries, they are still persistent where previously used, are still used elsewhere, and show up on foods. Others, such as organophosphates, tend to break down so rapidly that they are unlikely to show up on food unless applied to crops very close to harvest time, but they are often more acutely toxic. Approved pesticides leave little residue on crops when used according to directions. National and international standards of acceptable residue levels are based on approved usage and maximum acceptable daily intake (MADI) levels; thus, theoretically, foods should be safe. However, it is not practical to monitor all crops and shipments, residues are not always detected even when tests are done, and pesticides are not always used according to directions; thus, dangerous residues do make their way to the foods that people eat.

In the United States, public concern centers on the residues on imported foods

(Lecos, 1984). Although the Food and Drug Administration (FDA) maintains an active monitoring program that makes it advantageous for growers to comply with U.S. standards, only a small percentage of shipments can actually be tested. According to FDA figures, some 2000 food shipments crossing the Mexican border are tested annually (Thompson, 1984). This is indeed a small percentage of the Mexican import traffic, which exceeds 900 truckloads per day during the peak of the January-to-April shipping season. In 1986 the FDA checked nearly 8000 samples, less than 1 percent of the imported shipments of fruits and vegetables. Rather than hold perishable produce at border checkpoints for the average 18 days it takes for the test results, shipments are routinely allowed to pass (GAO, 1986b). Positive tests, indicating dangerous pesticide residues, result only in the stoppage of future shipments or in recalls, if the food can be traced. From 1979 to 1985 an average of 6 percent of the shipments sampled were found to have illegal residues—a rate about twice the average rate of contamination of domestic foods. Online little more than half (55 percent) of the illegal shipments were prevented from reaching consumers, and in only a few cases were damages collected from offending shippers (GAO, 1986a; Bull, 1982).

How often pesticide residues actually turn up on the table in the United States is largely unknown. The FDA's total-diet studies, conducted once a year in four regions of the country, reveal that pesticides do show up frequently but are "consistently" below the limits set by the Food and Agriculture Organization (FAG) and the World Health Organization (WHO) (Farley, 1988). DDE, a breakdown product of DOT, was still found in about 40 percent of the food sampled in 1983. This chemical, along with other organochlorines, has declined steadily in food since the 1970s. At the same time, organophosphates have increased, corresponding to their increased usage. Malathion was found in about 20 percent of the food sampled in 1983. Increases in pesticide residues found in potatoes and peanuts in the early 1980s stemmed from the greater use of sprout suppressants and soil fungicides, respectively (Penner, 1984). The EDB scare of the early 1980s is an example of a case where residues went undetected for a long time. EDB (ethylene dibromide) had been used since 1948 as a soil fumigant and as a postharvest fumigant on citrus and grains. Although a suspected carcinogen, EDB was thought to leave no residue, but the sensitive tests developed in the 1970s found that it did leave a residue. In 1983 it was found in cake mixes, cereals, and other foods from grains, and it was found in groundwater in several states where it had been used as a soil fumigant.

Public reaction was so strong that in September 1983 the EPA issued an emergency ban on the use of EDB as a soil fumigant. Bans on its use on grains and citrus followed in early 1984 ("Pesticide Is Banned," 1983; "New Moves," 1984; "EPA to Limit," 1984; Thompson, 1984). Although our food supply may now be nearly free of this dangerous chemical, over 30 years of use left a legacy of contaminated water and untold increases in cancer risk for the generation of consumers and farmers raised with it ("EDB's Long-Lasting Legacy," 1985; Barles and Kotas, 1987).

In third-world countries, pesticides are commonly misused because information on their proper use is not generally available. Their labeling is poor and often not in the native language, extension advisory services are usually unavailable, and proper usage outside of the temperate zone may be unknown, even by the manufacturers (Bull, 1982). In one market in India, a 7-year study showed that organochlorine residues were above the tolerance limits in more than 35 percent of the food. Forty percent of the samples of horticultural crops in Kenya contained excessive residues. A 1978 study. turned up rice samples from Sri Lanka that contained excessive residues of dieldrin and heptachlor. While exported foods are usually monitored carefully so that they will pass the inspections that the richer importing nations impose, locally marketed foods are usually not monitored, yet these are most likely to have residues of the short-lived, highly toxic organophosphates because of the short time between harvesting and marketing.

The responsibility for correcting this problem seems to lie with the richer nations of the world, which have profited by the export of pesticides and the technology of modern agriculture at the expense of the citizenry of third-world nations.

Pesticide Resistance

In addition to directly poisoning our environment and our food, pesticides pose a serious threat to our food-production system itself. From one viewpoint, pesticides are the wonder chemicals that have increased food production by 20 percent since 1940 by reducing pest damage. Yet over the same period they have also created at least 261 strains of insect species, 67 strains of plant pathogens, 2 strains of nematodes, and 4 (by some counts 19) strains of weeds that they cannot kill (Conway, 1982). While insecticide use increased tenfold since the 1940s, crop losses to insects doubled (Pimentel et al., 1983). The key to this paradox is the strong selection for resistance that pesticides exert on their target pests. Pesticides never kill 100 percent of a pest population, and the survivors tend to have a lower susceptibility to that particular chemical. With every repeated application of the same pesticide, these naturally resistant individuals make up a higher percentage of the population, until a highly resistant strain of pest evolves. When the conditions are right—the pesticide kills a large percentage of the pest population, the pest completes several life cycles per year, and little immigration from untreated populations occurs—resistance can develop very rapidly.

Resistance to one pesticide often confers resistance, or faster development of resistance, to a whole family of related pesticides (Bull, 1982). Alternating different pesticides or applying a mixture of chemicals can sometimes delay the development of resistance, but it can also promote the development of superresistant pests. Superpests, resistant to multiple pesticides, have already developed and threaten a number of crops throughout the world. So far, widespread famine has not resulted from superpests on food crops, but regional crises have occurred and some larger problems loom ahead. In North and Central America, cotton production has the worst record of superresistant pests. The cotton bollworm and budworm are both resistant to organochlorines, organophosphates, and carbamates and are becoming resistant to the new pyrethroids (FAO, 1979; Conway, 1982). Worldwide, 33 species of cotton pests were resistant to pesticides as of 1982, an increase from 20 in 1965. Rice, the staple food of a large proportion of the world population, now has 14 pest species resistant to one or more insecticides. Without pest control, experts estimate that 50 percent of the rice crop in Japan, India, and the Philippines would be lost to pests (Conway, 1982). The diamondback moth, which feeds on brassicaceous crops, has developed multiple resistance to at least 11 insecticides, including some from all the major groups, including pyrethroids (Bull, 1982). This pest now threatens the production of these crops in southeast Asia (FAG, 1979; Conway, 1982). Farmers there may spray three times a week and use 11 different insecticides in the course of a season, often spraying less than a week before harvest and in higher-than-recommended doses (Bull, 1982).

Pesticide resistance is considered inevitable, with no type of chemical immune (A. W. A. Brown, 1978); formerly effective pesticides cannot be successfully reintroduced (Bull, 1982). In order to avoid major pest-caused crop failures in the near future, we cannot simply rely on chemists developing new chemicals. Rotation of pesticides may delay the development of resistance in some cases, but these cases will require careful, regionally specific tailoring, and the research has yet to be done (Conway, 1982). Under normal field conditions, tolerating low kills slows the rate of resistance development because of the lower selection pressure exerted. Any increasing migration of pests from untreated areas into treated areas also slows the evolution of resistance (Conway, 1982). Both lower kills and higher migration probably mean that farmers would have to tolerate greater crop damage by pests. Integrated pest management is one approach to a longer-term solution, but it still may be vulnerable to the development of resistance. The development of pest-resistant crop strains, a major thrust of crop breeding programs, suffers much the same malady as the pesticide solution; it becomes a race between the breeder and the pest as to which will develop resistance first.

Rather than looking at ways to kill pests without exerting selection pressure for resistance, we might do better to turn the question around and ask, How can we successfully do agriculture without having to kill pests? Perhaps the cropping systems themselves can be altered to minimize pest population buildups in the first place. Traditional crop rotations and mixed-cropping systems hold potential for pest control by suppression rather than by killing (Pimentel, 1977). Development of new perennial crops, drawing on the greater natural resistance of perennials, holds great promise (Jackson, 1987).

Despite the above chronicle of ecological disasters that attend it, agriculture is still potentially a renewable enterprise. In a certain sense, nothing is new here. In every century for 10,000 years on almost every agricultural acre in the world, ecological capital has slipped seaward from its wilderness-built home or been degraded by toxic materials. Still, on a global scale, agriculture is seen as potentially renewable and therefore as fundamentally different from the industrial sector of society. It is only in the last 50 years, with the expansion of industry and the chemicalization of agriculture, that the inherently extractive economy has acted as though the renewable resources that support agriculture are fair targets for exploitation in industrial terms. That is what makes the modern era different. That is what makes the current agricultural economy more brittle than almost any agricultural economy in history. It is hard for us to see this, perhaps, because it is hard for us to imagine our energy and aquifer mines lying hollow. It is hard for us to believe that our well water and air, our pleasingly packaged food, and our perfect produce could contain invisible poisons. It is hard for us to grasp the value of genetic diversity or to foresee the consequences of rampant extinctions. It is hard for us to believe that we may one day come to the end of our magical ability to produce everhigher-yielding crops. It is hard for us to comprehend the total loss of our vast tropical forests or to anticipate the climatic changes when the earth's belly is belted in barren, sterile soils rather than the green, moist vegetation. It is hard for us to imagine a world where ancient salts sterilize the land and young chemical pesticides and fertilizers lie below our agricultural surfaces like demons in quiet prisons of degraded soil.

The outlook is not entirely bleak, for solutions to all these problems lie in lessons we have learned and can still learn from nature. If we turn our attention away from the extractive industrial model and begin to focus on nature's models of productive ecosystems as our guide for agricultural systems, we may yet see truly sustainable agriculture emerging. It isn't that nature learns faster than humans. It is just that she has been at it longer.

The Myths of Agricultural Biotechnology: Some Ethical Questions

For years academicians have assumed that agriculture poses no special problem for environmental ethics, despite the fact that human life and human civilization depend on the intentional artificialization of nature to carry out agricultural production. Even critics of the environmental impacts of pesticides and of the social implications of agricultural technology have failed to conceptualize a coherent environmental ethics applicable to agricultural problems (Thompson 1995). In general, most Proponents of sustainable agriculture, driven by a technological determinism lack on understanding of the structural roots of the environmental degradation linked to capitalist agriculture. Therefore, by accepting the present socioeconomic and political structure of agriculture as a given, they became constrained from implementing an alternative agriculture that challenges such structure (Levins and Lewontin 1985). This is worrisome, especially today, as profit motivations rather than environmental concerns, shape the type of research and modes of agricultural production prevalent throughout the world (Busch et al. 1990).

Herein we contend that the key problem facing agroecologists, is that modern industrial agriculture, today epitomized by biotechnology, is founded on philosophical premises that are fundamentally flawed, and that precisely those premises are the ones that need to be exposed and criticized in order to advance towards a truly sustainable agriculture.

This is particularly relevant in the case of biotechnology, where the alliance of reductionist science and a multinational monopolistic industry which jointly perceive agricultural problems as genetic deficiencies of organisms and treat nature as a commodity, will take agriculture further down a misguided route (Levidow and Carr 1997).

The objective of this paper is to challenge the false promises made by the genetic engineering industry that it will move agriculture away from a dependence on chemical inputs, that will increase productivity, as well as decrease input costs and help reduce environmental problems (OTA 1992). By challenging the myths of biotechnology, we expose genetic engineering for what it really is; another "technological fix" or "magic bullet" aimed at circumventing the environmental problems of agriculture (which themselves are the outcome of and earlier round of technological fix), without questioning the flawed assumptions that gave rise to the problems in the first place (Hindmarsh 1991). Biotechnology develops single-gene solutions for problems that derive from ecologically unstable monoculture systems, designed on industrial models of efficiency. Such unilateral approach was already proven ecologically unfit in the case of pesticides (Pimentel et al. 1992).

Ethical Questions About Biotechnology

Environmentalists critical of biotechnology, question the assumptions that biotechnological science is value free, and that it cannot be wrong or misused and call for an ethical evaluation of genetic engineering research and its products (Krimsky and Wrubel 1996). Proponents of biotechnology are perceived as having a utilitarian view of nature and as favoring the free trading of economic gains for ecological damage with indifference to the human consequences (James 1997). At the very heart of the critique are biotechnology's effects on social and economic conditions and religious and moral values giving rise to questions such as:

Should we alter the genetic structure of the entire living kingdom in the name of utility and profit?

Is there something sacred about life, or should life forms, including humans, be viewed simply as commodities in the new biotechnological marketplace?

Is the genetic makeup of all living things, the common heritage of all, or it can be appropriated by corporations and thus become private property of a few?

Who gave individual companies the right to the monopoly over entire groups of organisms?

Do biotechnologists feel as masters of nature? Is this an illusion constructed on scientific arrogance and conventional economics, blind to the complexity of ecological processes?

It is possible to minimize ethical concerns and reduce environmental risks while keeping the benefits?

There are also questions that arise specifically from the nature of the technology, while others such as the domination of agricultural research agendas by commercial interests, the uneven distribution of benefits, the possible environmental risks and the exploitation of the poor nations' genetic resources by rich ones demand a deeper inquiry:

- Who benefits from the technology? Who losses?
- What are the environmental and health consequences?
- What have been the alternatives forgone?
- To whose needs does biotechnology respond?
- How does the technology affect what is being produced, how it is being produced and for what and for whom?
- What are the social goals and ethical criteria that guides research problem choices?
- Biotechnology for achieving what social and agronomic goals?

The Biotechnology Myths

The agrochemical corporations which control the direction and goals of agricultural innovation through biotechnology claim that genetic engineering will enhance the sustainability of agriculture by solving the very problems affecting conventional farming and will spare Third World farmers from low productivity, poverty and hunger (Molnar and Kinnucan 1989, Gresshoft 1996). By matching myth with reality the following section describes how and why current developments in agricultural biotechnology do not measure up to such promises and expectations.

Myth 1: Biotechnology will benefit farmers in the US and in the developed world.

Most innovations in agricultural biotechnology are profit driven rather than need driven, therefore the thrust of the genetic engineering industry is not to solve agricultural problems as much as it is to create profitability. Moreover, biotechnology seeks to industrialize agriculture even further and to intensify farmers' dependence upon industrial inputs aided by a ruthless system of intellectual property rights which legally inhibits the right of farmers to reproduce, share and store seeds (Busch et al. 1990). By controlling the germplasm from seed to sale and by forcing farmers to pay inflated prices for seed-chemical packages, companies are determined to extract the most profit from their investment.

Because biotechnologies are capital intensive they will continue to deepen the pattern of change in US agriculture, increasing concentration of agricultural production in the hands of large-corporate farms. As with other labor saving technology, by increasing productivity biotechnology tends to reduce commodity prices and set in motion a technology treadmill that forces out of business a significant number of farmers, especially small scale. The example of bovine growth hormone confirms the hypothesis that biotechnology will accelerate the foreclosure of small dairy farms (Krimsky and Wrubel 1996).

Myth 2: Biotechnology will benefit small farmers and will favor the hungry and poor of the Third World.

If Green Revolution technology bypassed small and resource-poor farmers, biotechnology will exacerbate marginalization even more as such technologies are under corporate control and protected by patents, are expensive and inappropriate to the needs and circumstances of indigenous people (Lipton 1989). As biotechnology is primarily a commercial activity, this reality determines priorities of what is investigated, how it is applied and who is to benefit. While the world may lack food and suffer from pesticide pollution, the focus of multinational corporations is profit, not philanthropy. This is why biotechnologists design transgenic crops for new marketable quality or for import substitution, rather than for greater food production (Mander and Goldsmith 1996). In general, biotechnology companies are emphasizing a limited range of crops for which there are large and secured markets, targeted at relatively capital-intensive production systems. As transgenic crops are patented plants, this means that indigenous farmers can lose rights to their own regional germplasm and not be allowed under GATT to reproduce, share or store the seeds of their harvest (Crucible Group 1994). It is difficult to conceive how such technology will be introduced in Third World countries to favor the masses of poor farmers. If biotechnologists were really committed to feeding the world, why isn't the scientific genius of biotechnology turned to develop varieties of crops more tolerant to weeds rather than to herbicides? Or why aren't more promising products of biotechnology, such as N fixing and drought tolerant plants being developed?

Biotechnology products will undermine exports from the Third World countries especially from small-scale producers. The development of a thaumatin product via biotechnology is just the beginning of a transition to alternative sweeteners which will replace Third World sugar markets in the future (Mander and Goldsmith 1996). It is estimated that nearly 10 million sugar farmers in the Third World may face a loss of livelihood as laboratory-processed sweeteners begin invading world markets. Fructose produced by biotechnology already captured over 10% of the world market and caused sugar prices to fall, throwing tens of thousands of workers out of work. But such foreclosures of rural opportunities are not limited to sweeteners. Approximately 70,000 vanilla farmers in Madagascar were ruined when a Texas firm produced vanilla in biotech labs (Busch et al. 1990). The expansion on Unilever cloned oil palms will substantially increase palm-oil production with dramatic consequences for farmers producing other vegetable oils (groundnut in Senegal and coconut in Philippines).

Myth 3: Biotechnology will not attempt against the ecological sovereignty of the Third World.

Ever since the North became aware of the ecological services performed by biodiversity of which the South is the major repository, the Third World has witnessed a "gene rush" as multinational corporations aggressively scour forests, crop fields and coasts in search of the South's genetic gold (Kloppenburg 1988). Protected by GATT, MNCs freely practice "biopiracy" which the Rural Advancement Foundation (RAFI) estimates it costing developing countries US $ 5.4 billion a year through lost royalties from food and drug companies which use indigenous farmers' germplasm and medicinal plants (Levidow and Carr 1997).

Clearly, indigenous people and their biodiversity are viewed as raw materials for the MNCs which have made billions of dollars on seeds developed in US labs from germplasm that farmers in the Third World had carefully bred over generations (Fowler and Mooney 1990). Meanwhile, peasant farmers go unrewarded for their millenary farming knowledge, while MNCs stand to harvest royalties from Third World countries estimated at billions of dollars. So far biotechnology companies offer no provisions to pay Third World farmers for the seeds they take and use (Kloppenburg 1988).

Myth 4: Biotechnology will lead to biodiversity conservation.

Although biotechnology has the capacity to create a greater variety of commercial plants and thus contribute to biodiversity, this is unlikely to happen. The strategy of MNCs is to create broad international seed markets for a single product. The tendency is towards uniform international seed markets (MacDonald 1991). Moreover, the MNC-dictated provisions of the patent system prohibiting farmers to reuse the seed yielded by their harvests, will affect the possibilities of in-situ conservation and on-farm improvements of genetic diversity.

The agricultural systems developed with transgenic crops will favor monocultures characterized by dangerously high levels of genetic homogeneity leading to higher vulnerability of agricultural systems to biotic and abiotic stresses (Robinson 1996). As the new bioengineered seeds replace the old traditional varieties and their wild relatives, genetic erosion will accelerate in the Third World (Fowler and Mooney 1990). Thus the push for uniformity will not only destroy the diversity of genetic resources, but will also disrupt the biological complexity that underlines the sustainability of traditional farming systems (Altieri 1994).

Myth 5: Biotechnology is ecologically safe and will launch a period of a chemical-free sustainable agriculture.

Biotechnology is being pursued to patch-up the problems that have been caused by previous agrochemical technologies (pesticide resistance, pollution, soil degradation, etc.) which were promoted by the same companies now leading the bio-revolution. Transgenic crops developed for pest control follow closely the pesticide paradigm of using a single control mechanism which has proven to fail over and over again with insects, pathogens and weeds (NRC 1996). Transgenic crops are likely to increase the use of pesticides and to accelerate the evolution of "super weeds" and resistant insect pests strains (Rissler and Mellon 1996). The "one gene — one pest" resistant approach has proven to be easily overcome by pests which are continuously adapting to new situations and evolving detoxification mechanisms (Robinson 1997).

There are many unanswered ecological questions regarding the impact of the release of transgenic plants and micro-organisms into the environment.

Among the major environmental risks associated with genetically engineered plants are the unintended transfer to plant relatives of the "transgenes" and the unpredictable ecological effects (Rissler and Mellon 1996).

Given the above considerations, agroecological theory predicts that biotechnology will exacerbate the problems of conventional agriculture and by promoting monocultures will also undermine ecological methods of farming such as rotation and polycultures (Hindmarsh 1991). As presently conceived, biotechnology does not fit into the broad ideals of a sustainable agriculture (Kloppenburg and Burrows 1996).

Myth 6: Biotechnology will enhance the use of molecular biology for the benefit of all sectors of society.

The demand for the new biotechnology did not emerge as a result of social demands but it emerged out of changes in patent laws and the profit interests of chemical companies of linking seeds and pesticides. The supply emerged out of breakthroughs in molecular biology and the availability of venture capital as a result of favorable tax laws (Webber 1990). The danger is that the private sector is influencing the direction of public sector research in ways unprecedented in the past (Kleinman and Kloppenburg 1988).

As more universities enter into partnerships with corporations, serious ethical questions emerge about who owns the results of research and which research gets done. The trend toward secrecy by university scientists involved in such partnerships raises questions about personal ethics and conflicts of interest. In many universities a professor's ability to attract private investment is often more important than academic qualifications, taking away the incentives for scientists to be socially responsible. Fields such as biological control and agroecology which do not attract corporate sponsorship are being phased out and this not in the public interest (Kleinman and Koppenburg 1988).

In the late 1980's, a statement issued by Monsanto indicated that biotechnology would revolutionize agriculture in the future with products based on nature's own methods, making farming more environmentally friendly and more profitable for the farmer (OTA 1992). Moreover, plants would be provided with built-in defenses against insects and pathogens. Since then many others have promised several more valuable rewards that biotechnology can bring through crop improvement. The ethical dilemma is that many of these promises are unfounded and many of the advantages or benefits of biotechnology have not or may not be realized. Although clearly biotechnology holds promise for an improved agriculture, given its present orientation it mostly holds promise for environmental harm, for the further industrialization of agriculture and for the intrusion of private interests too far into public interest sector research. Until now, the economic and political domination of the agricultural development agenda by MNCs has thriven at the expense of the interests of consumers, farm workers, small family farms, wildlife and the environment.

It is urgent for civil society to have earlier entry points and broader participation in technological decisions so that the domination of scientific research by corporate interests is dealt with more stringent public control. National and international public organizations such as FAO, CGIAR, etc., will have to carefully monitor and control the provision of applied non proprietary knowledge to the private sector so as to protect that such knowledge will continue in the public domain for the benefit of rural societies. Publicly controlled regulatory regimes must be developed and employed for assessing and monitoring the environmental and social risks of biotechnological products (Webber 1990).

Finally, the trends towards a reductionist view of nature and agriculture set in motion by contemporary biotechnology must be reversed by a more holistic approach to agriculture, so as to ensure that agroecological alternatives are not foregone and that only ecologically-sound aspects of biotechnology are researched and developed. The time has come to counter effectively the challenge, and the reality, of genetic engineering. As it has been with pesticides, biotechnology companies must feel the impact of environmental, farm labor, animal rights and consumers lobbies, so that they start re-orienting their work for the overall benefit of society and nature. The future of biotechnology based research will be determined by power relations, and there is no reason why farmers and the public in general, if sufficiently empowered, could not influence the direction of biotechnology along sustainability goals.

9

Global Agricultural Agendas & Policy Recommendations

Agricultural Science Policy Changing Global Agendas

In the second half of the 20th century, agricultural research and development (R&D) generated astounding increases in food production. Technological advances devel- oped through R&D have supplied the world with not only *more* food, but *better* food. Since the 1980s, however, the growth of public spending on agricultural R&D has slowed dramatically in both developed and developing countries, while the need to pro- duce enough food to feed burgeoning populations grows more pressing with every pass- ing day. About 800 million people are undernourished today. Over the next 25 years, demand for cereals is expected to grow by over 40 percent; for meat, 70 percent To achieve the necessary yield growth without harming the environment is an enormous chal- lenge. This challenge will have to be met largely through agricultural research under the shadow of ever-tighter public research budgets.

In conjunction with budgetary pressures, the emphasis of spending is shifting away from research on farm inputs and outputs toward new concerns such as the environment, genetic diversity, food safety, and human health. Research and technology policy is also being made more complicated by the changing public and private roles in funding and conducting agricultural R&D, changing intellectual property rights regimes and changes to the general environment for products and trade in agricultural products *Agricultural Science Policy: Changing Global Agendas*, edited by Julian M. Alston Philip G. Pardey, and Michael J. Taylor, and published by Johns Hopkins University Press, looks at issues raised by this changing environment for agricultural productivity agricultural R&D, and natural resource management.

Changing Contexts and Agents

Today, private entities are paying for and conducting a larger share of research while many governments are scaling back the growth in or reducing their spending on agri- cultural R&D. Spending on private agricultural R&D in the rich countries alone totaled US$7 billion in 1993, compared with US$4 billion in 1981. Changes in the world econo- my and in agriculture and science are having important implications for the funding, organization, and management of R&D. In many countries the emergence of modern biotechnology and a strengthening of intellectual property rights are giving rise to pro- found changes in both public and private agricultural R&D. Only a small part (12 percent) of private research concerns farm-level technologies for improving crop and livestock production. Much research relates to technologies embodied in purchased inputs as well as post-farm food processing and product development. Specialization has become the name of the game: in one country, firms may emphasize postharvest research; in another, pesticides and herbicides. Research results tend to be rapidly transferred around the world.

In this changing research environment, a number of important policy questions must be addressed. How should resources for agricultural research be distributed among local, national, and international research? How much agricultural research should be publicly funded? What role should multinational firms play? How should scarce research dollars be distributed among sectors such as agriculture, forestry, fisheries, and livestock? Should a large part of agricultural research be directed toward environmental issues such as cli- mate change, loss of biodiversity, and water pollution? Should research concentrate on short-term or long-term results? On improving the environment or increasing yields?

This book lays out this new agenda for agricultural science policy and provides new evidence and new ideas as a basis for beginning to study and formulate policies to deal with the emerging issues.

Productivity Measures and Measurement Challenges

Setting research priorities requires accurately measuring the effects of alternative courses of action, a task that raises conceptual and measure- ment issues. One, for example, emphasizes the effects of adjusting for changes in the quality of outputs and inputs (like aging but better educated farmers or bigger tractors and combine har- vesters) in measures of state-level productivity patterns in the United States. Another study looks at past trends and future prospects for cereal crop productivity as growth in rice and wheat yields slows in developing countries. Yield is still the most com- monly used indicator of productivity, but proper assessment of productiv- ity growth also requires that attention be paid to trends in the use of other factors of production such as fertilizer. Empirical studies of total factor pro- ductivity—measuring the output not attributable to the inputs used in agri- culture—in developing-country agri- culture are becoming increasingly important in providing a complete picture of technological change.

Research, Productivity, and Natural Resources

Can research and technology solve environmental problems of water, soil, and air degradation, without adversely affecting economic develop- ment and agricultural productivity? What are the implications of agricul- tural R&D for sustainability? The very notion of sustainability, as well as other popular ideas about the nature and extent of resource scarcity, includ- ing the effects of agriculture on natural resource stocks and the implications for science and related policy. Reg- ulated industries based on nonrenew- able or common property resources raise new issues in the measurement of productivity, as illustrated by an analy- sis of technical change and productivity in the North Pacific halibut fishery during 1935–78. Getting the facts right is important as well. New data on soil degradation in China and Indonesia show that, on balance, over very long time periods the contributions to soil fertility associated with urban devel- opment and intensification can more than offset the losses from land degradation in the hinterlands.

Research for Genetic Improvement

Studies of genetic improvements in crops have dominated the large body of work on the economics of agricultural R&D. This research is evolving to address new types of issues. For instance, this study found that international wheat and maize research and Indian wheat research have generated significant spillovers (in which research conducted in one location affects a different location) and that agricultural research has large economies of size, scale, and scope. Policies related to the conservation of agricultural plant biodiversity are addressed in a study that discusses methods for valuing genetic resources and the implications of property rights regimes for plant breeding programs. Another area of concern today is food safety, particularly in transgenic crops, but genetically modified crops can also promote human health. For instance, the health benefits from consuming canola oil are reported to be as large as the gross value of the crop, much greater than the benefits to producers and consumers measured conventionally.

AGENDA 21 (CHAPTER 14): PROMOTING SUSTAINABLE AGRICULTURE AND RURAL DEVELOPMENT

1. By the year 2025, 83 per cent of the expected global population of 8.5 billion will be living in developing countries. Yet the capacity of available resources and technologies to satisfy the demands of this growing population for food and other agricultural commodities remains uncertain. Agriculture has to meet this challenge, mainly by increasing production on land already in use and by avoiding further encroachment on land that is only marginally suitable for cultivation.

2. Major adjustments are needed in agricultural, environmental and macroeconomic

policy, at both national and international levels, in developed as well as developing countries, to create the conditions for sustainable agriculture and rural development (SARD). The major objective of SARD is to increase food production in a sustainable way and enhance food security. This will involve education initiatives, utilization of economic incentives and the development of appropriate and new technologies, thus ensuring stable supplies of nutritionally adequate food, access to those supplies by vulnerable groups, and production for markets; employment and income generation to alleviate poverty; and natural resource management and environmental protection.

3. The priority must be on maintaining and improving the capacity of the higher potential agricultural lands to support an expanding population. However, conserving and rehabilitating the natural resources on lower potential lands in order to maintain sustainable man/land ratios is also necessary. The main tools of SARD are policy and agrarian reform, participation, income diversification, land conservation and improved management of inputs. The success of SARD will depend largely on the support and participation of rural people, national Governments, the private sector and international cooperation, including technical and scientific cooperation.

4. The following programme areas are included in this study:

a. Agricultural policy review, planning and integrated programming in the light of the multifunctional aspect of agriculture, particularly with regard to food security and sustainable development;

b. Ensuring people's participation and promoting human resource development for sustainable agriculture;

c. Improving farm production and farming systems through diversification of farm and non-farm employment and infrastructure development;

d. Land-resource planning information and education for agriculture;

e. Land conservation and rehabilitation;

f. Water for sustainable food production and sustainable rural development;

g. Conservation and sustainable utilization of plant genetic resources for food and sustainable agriculture;

h. Conservation and sustainable utilization of animal genetic resources for sustainable agriculture;

i. Integrated pest management and control in agriculture;

j. Sustainable plant nutrition to increase food production;

k. Rural energy transition to enhance productivity;

l. Evaluation of the effects of ultraviolet radiation on plants and animals caused by the depletion of the stratospheric ozone layer.

Programme Areas

A. Agricultural Policy Review, Planning and Integrated Programmes in the Light of the Multifunctional Aspect of Agriculture, Particularly with Regard to Food Security and Sustainable Development

Basis for Action

5. There is a need to integrate sustainable development considerations with agricultural policy analysis and planning in all countries, particularly in developing countries. Recommendations should contribute directly to development of realistic and operational medium- to long-term plans and programmes, and thus to concrete actions. Support to and monitoring of implementation should follow.

6. The absence of a coherent national policy framework for sustainable agriculture and rural development (SARD) is widespread and is not limited to the developing countries. In particular the economies in transition from planned to market-oriented systems need such a framework to incorporate environmental considerations into economic activities, including agriculture. All countries need to assess comprehensively the impacts of such policies on food and agriculture sector performance, food security, rural welfare and international trading relations as a means for identifying appropriate offsetting measures. The major thrust of food security in this case is to bring about a significant increase in agricultural production in a sustainable way and to achieve a substantial improvement in people's entitlement to adequate food and culturally appropriate food supplies.

7. Sound policy decisions pertaining to international trade and capital flows also necessitate action to overcome: (a) a lack of awareness of the environmental costs incurred by sectoral and macroeconomic policies and hence their threat to sustainability; (b) insufficient skills and experience in incorporating issues of sustainability into policies and programmes; and (c) inadequacy of tools of analysis and monitoring.1

Objectives

8. The objectives of this Programme area are:

a. By 1995, to review and, where appropriate, establish a programme to integrate environmental and sustainable development with policy analysis for the food and agriculture sector and relevant macroeconomic policy analysis, formulation and implementation;

b. To maintain and develop, as appropriate, operational multisectoral plans, programmes and policy measures, including programmes and measures to enhance sustainable food production and food security within the framework of sustainable development, not later than 1998;

c. To maintain and enhance the ability of developing countries, particularly the least developed ones, to themselves manage policy, programming and planning activities, not later than 2005.

Activities

(a) Management-related Activities

9. Governments at the appropriate level, with the support of the relevant international and regional organizations, should:

a. Carry out national policy reviews related to food security, including adequate levels and stability of food supply and access to food by all households;

b. Review national and regional agricultural policy in relation, inter alia, to foreign trade, price policy, exchange rate policies, agricultural subsidies and taxes, as well as organization for regional economic integration;

c. Implement policies to influence land tenure and property rights positively with due recognition of the minimum size of land-holding required to maintain production and check further fragmentation;

d. Consider demographic trends and population movements and identify critical areas for agricultural production;

e. Formulate, introduce and monitor policies, laws and regulations and incentives leading to sustainable agricultural and rural development and improved food security and to the development and transfer of appropriate farm technologies, including, where appropriate, low-input sustainable agricultural (LISA) systems;

f. Support national and regional early warning systems through food-security assistance schemes that monitor food supply and demand and factors affecting household access to food;

g. Review policies with respect to improving harvesting, storage, processing, distribution and marketing of products at the local, national and regional levels;

h. Formulate and implement integrated agricultural projects that include other natural resource activities, such as management of rangelands, forests, and wildlife, as appropriate;

i. Promote social and economic research and policies that encourage sustainable agriculture development, particularly in fragile ecosystems and densely populated areas;

j. Identify storage and distribution problems affecting food availability; support research, where necessary, to overcome these problems and cooperate with producers and distributors to implement improved practices and systems.

(b) Data and Information

10. Governments at the appropriate level, with the support of the relevant international and regional organizations, should:

a. Cooperate actively to expand and improve the information on early warning systems on food and agriculture at both regional and national levels;

b. Examine and undertake surveys and research to establish baseline information on the status of natural resources relating to food and agricultural production and planning in order to assess the impacts of various uses on these resources, and develop methodologies and tools of analysis, such as environmental accounting.

(c) International and Regional Cooperation and Coordination

11. United Nations agencies, such as FAO, the World Bank, IFAD and GATT, and regional organizations, bilateral donor agencies and other bodies should, within their respective mandates, assume a role in working with national Governments in the following activities:

a. Implement integrated and sustainable agricultural development and food security strategies at the subregional level that use regional production and trade potentials, including organizations for regional economic integration, to promote food security;

b. Encourage, in the context of achieving sustainable agricultural development and consistent with relevant internationally agreed principles on trade and environment, a more open and non-discriminatory trading system and the avoidance of unjustifiable trade barriers which together with other policies will facilitate the further integration of agricultural and environmental policies so as to make them mutually supportive;

c. Strengthen and establish national, regional and international systems and networks to increase the understanding of the interaction between agriculture and the state of the environment, identify ecologically sound technologies and facilitate the exchange information on data sources, policies, and techniques and tools of analysis.

Means of Implementation

(a) Financing and Cost Evaluation

12. The Conference secretariat has estimated the average total annual cost (1993-2000) on implementing the activities of this programme to be about $3 billion, including about $450 million from the international community on grant or concessional terms. These are indicative and order-of-magnitude estimates only and have not been reviewed

by Governments. Actual costs and financial terms, including any that are non-concessional, will depend upon, inter alia, the specific strategies and programmes Governments decide upon for implementation.

(b) Scientific and Technological Means

13. Governments at the appropriate level and with the support of the relevant international and regional organizations should assist farming households and communities to apply technologies related to improved food production and security, including storage, monitoring of production and distribution.

(c) Human Resource Development

14. Governments at the appropriate level, with the support of the relevant international and regional organizations, should:

a. Involve and train local economists, planners and analysts to initiate national and international policy reviews and develop frameworks for sustainable agriculture;

b. Establish legal measures to promote access of women to land and remove biases in their involvement in rural development.

(d) Capacity-building

15. Governments at the appropriate level, with the support of the relevant international and regional organizations, should strengthen ministries for agriculture, natural resources and planning.

B. Ensuring People's Participation and Promoting Human Resource Development for Sustainable Agriculture

Basis for Action

16. This component bridges policy and integrated resource management. The greater the degree of community control over the resources on which it relies, the greater will be the incentive for economic and human resources development. At the same time, policy instruments to reconcile long-run and short-run requirements must be set by national Governments. The approaches focus on fostering self-reliance and cooperation, providing information and supporting user-based organizations. Emphasis should be on management practices, building agreements for changes in resource utilization, the rights and duties associated with use of land, water and forests, the functioning of markets, prices, and the access to information, capital and inputs. This would require training and capacity-building to assume greater responsibilities in sustainable development efforts.

Objectives

17. The objectives of this programme area are:

a. To promote greater public awareness of the role of people's participation and people's organizations, especially women's groups, youth, indigenous people, local communities and small farmers, in sustainable agriculture and rural development;

b. To ensure equitable access of rural people, particularly women, small farmers, landless and indigenous people, to land, water and forest resources and to technologies, financing, marketing, processing and distribution;

c. To strengthen and develop the management and the internal capacities of rural people's organizations and extension services and to decentralize decision-making to the lowest community level.

Activities

(a) Management-related activities

18. Governments at the appropriate level, with the support of the relevant international and regional organizations, should:

a. Develop and improve integrated agricultural extension services and facilities and rural organizations and undertake natural resource management and food security activities, taking into account the different needs of subsistence agriculture as well as market-oriented crops;

b. Review and refocus existing measures to achieve wider access to land, water and forest resources and ensure equal rights of women and other disadvantaged groups, with particular emphasis on rural populations, indigenous people and local communities;

c. Assign clear titles, rights and responsibilities for land and for individuals or communities to encourage investment in land resources;

d. Develop guidelines for decentralization policies for rural development through reorganization and strengthening of rural institutions;

e. Develop policies in extension, training, pricing, input distribution, credit and taxation to ensure necessary incentives and equitable access by the poor to production-support services;

f. Provide support services and training, recognizing the variation in agricultural circumstances and practices by location; the optimal use of on-farm inputs and the minimal use of external inputs; optimal use of local natural resources and management of renewable energy sources; and the establishment of networks that deal with the exchange of information on alternative forms of agriculture.

(b) Data and information

19. Governments at the appropriate level, and with the support of the relevant international and regional organizations, should collect, analyse, and disseminate information on human resources, the role of Governments, local communities and non-governmental organizations in social innovation and strategies for rural development.

(c) International and regional cooperation and coordination

20. Appropriate international and regional agencies should:

a. Reinforce their work with non-governmental organizations in collecting and disseminating information on people's participation and people's organizations, testing participatory development methods, training and education for human resource development and strengthening the management structures of rural organizations;

b. Help develop information available through non-governmental organizations and promote an international ecological agricultural network to accelerate the development and implementation of ecological agriculture practices.

Means of implementation

(a) Financing and cost evaluation

21. The Conference secretariat has estimated the average total annual cost (1993-2000) of implementing the activities of this programme to be about $4.4 billion, including about $650 million from the international community on grant or concessional terms. These are indicative and order-of-magnitude estimates only and have not been reviewed by Governments. Actual costs and financial terms, including any that are non-concessional, will depend upon, inter alia, the specific strategies and programmes Governments decide upon for implementation.

(b) Scientific and technological means

22. Governments at the appropriate level, with the support of the relevant international and regional organizations, should:

a. Encourage people's participation on farm technology development and transfer, incorporating indigenous ecological knowledge and practices;

b. Launch applied research on participatory methodologies, management strategies and local organizations.

(c) Human resource development

23. Governments at the appropriate level, with the support of the relevant international and regional organizations, should provide management and technical

training to government administrators and members of resource-user groups in the principles, practice and benefits of people's participation in rural development.

(d) Capacity-building

24. Governments at the appropriate level, with the support of the relevant international and regional organizations, should introduce management strategies and mechanisms, such as accounting and audit services for rural people's organizations and institutions for human resource development, and delegate administrative and financial responsibilities to local levels for decision-making, revenue-raising and expenditure.

Basis for Action

25. Agriculture needs to be intensified to meet future demands for commodities and to avoid further expansion onto marginal lands and encroachment on fragile ecosystems. Increased use of external inputs and development of specialized production and farming systems tend to increase vulnerability to environmental stresses and market fluctuations. There is, therefore, a need to intensify agriculture by diversifying the production systems for maximum efficiency in the utilization of local resources, while minimizing environmental and economic risks. Where intensification of farming systems is not possible, other on-farm and off-farm employment opportunities should be identified and developed, such as cottage industries, wildlife utilization, aquaculture and fisheries, non-farm activities, such as light village-based manufacturing, farm commodity processing, agribusiness, recreation and tourism, etc.

Objectives

26. The objectives of this programme area are:

a. To improve farm productivity in a sustainable manner, as well as to increase diversification, efficiency, food security and rural incomes, while ensuring that risks to the ecosystem are minimized;

b. To enhance the self-reliance of farmers in developing and improving rural infrastructure, and to facilitate the transfer of environmentally sound technologies for integrated production and farming systems, including indigenous technologies and the sustainable use of biological and ecological processes, including agroforestry, sustainable wildlife conservation and management, aquaculture, inland fisheries and animal husbandry;

c. To create farm and non-farm employment opportunities, particularly among the poor and those living in marginal areas, taking into account the alternative livelihood proposal inter alia in dryland areas.

Activities

(a) Management-related Activities

27. Governments at the appropriate level, with the support of the relevant international and regional organizations, should:

a. Develop and disseminate to farming households integrated farm management technologies, such as crop rotation, organic manuring and other techniques involving reduced use of agricultural chemicals, multiple techniques for sources of nutrients and the efficient utilization of external inputs, while enhancing techniques for waste and by-product utilization and prevention of pre- and post-harvest losses, taking particular note of the role of women;

b. Create non-farm employment opportunities through private small-scale agro-processing units, rural service centres and related infrastructural improvements;

c. Promote and improve rural financial networks that utilize investment capital resources raised locally;

d. Provide the essential rural infrastructure for access to agricultural inputs and services, as well as to national and local markets, and reduce food losses;

e. Initiate and maintain farm surveys, on-farm testing of appropriate technologies and dialogue with rural communities to identify constraints and bottlenecks and find solutions;

f. Analyse and identify possibilities for economic integration of agricultural and forestry activities, as well as water and fisheries, and to take effective measures to encourage forest management and growing of trees by farmers (farm forestry) as an option for resource development.

(b) Data and Information

28. Governments at the appropriate level, with the support of the relevant international and regional organizations, should:

a. Analyse the effects of technical innovations and incentives on farm-household income and well-being;

b. Initiate and maintain on-farm and off-farm programmes to collect and record indigenous knowledge.

(c) International and Regional Cooperation and Coordination

29. International institutions, such as FAO and IFAD, international agricultural research centres, such as CGIAR, and regional centres should diagnose the world's

major agro-ecosystems, their extension, ecological and socio-economic characteristics, their susceptibility to deterioration and their productive potential. This could form the basis for technology development and exchange and for regional research collaboration.

Means of Implementation

(a) Financing and Cost Evaluation

30. The Conference secretariat has estimated the average total annual cost (1993-2000) of implementing the activities of this programme to be about $10 billion, including about $1.5 billion from the international community on grant or concessional terms. These are indicative and order-of-magnitude estimates only and have not been reviewed by Governments. Actual costs and financial terms, including any that are non-concessional, will depend upon, inter alia, the specific strategies and programmes Governments decide upon for implementation.

(b) Scientific and Technological Means

31. Governments at the appropriate level, with the support of the relevant international and regional organizations, should strengthen research on agricultural production systems in areas with different endowments and agro-ecological zones, including comparative analysis of the intensification, diversification and different levels of external and internal inputs.

(c) Human Resource Development

32. Governments at the appropriate level, with the support of the relevant international and regional organizations, should:

a. Promote educational and vocational training for farmers and rural communities through formal and non-formal education;

b. Launch awareness and training programmes for entrepreneurs, managers, bankers and traders in rural servicing and small-scale agro-processing techniques.

(d) Capacity-building

33. Governments at the appropriate level, with the support of the relevant international and regional organizations, should:

a. Improve their organizational capacity to deal with issues related to off-farm activities and rural industry development;

b. Expand credit facilities and rural infrastructure related to processing, transportation and marketing.

D. Land-resource Planning, Information and Education for Agriculture

Basis for Action

34. Inappropriate and uncontrolled land uses are a major cause of degradation and depletion of land resources. Present land use often disregards the actual potentials, carrying capacities and limitations of land resources, as well as their diversity in space. It is estimated that the world's population, now at 5.4 billion, will be 6.25 billion by the turn of the century. The need to increase food production to meet the expanding needs of the population will put enormous pressure on all natural resources, including land.

35. Poverty and malnutrition are already endemic in many regions. The destruction and degradation of agricultural and environmental resources is a major issue. Techniques for increasing production and conserving soil and water resources are already available but are not widely or systematically applied. A systematic approach is needed for identifying land uses and production systems that are sustainable in each land and climate zone, including the economic, social and institutional mechanisms necessary for their implementation. 3

Objectives

36. The objectives of this programme area are:

a. To harmonize planning procedures, involve farmers in the planning process, collect land-resource data, design and establish databases, define land areas of similar capability, identify resource problems and values that need to be taken into account to establish mechanisms to encourage efficient and environmentally sound use of resources;

b. To establish agricultural planning bodies at national and local levels to decide priorities, channel resources and implement programmes.

Activities

(a) Management-related Activities

37. Governments at the appropriate level, with the support of the relevant international and regional organizations, should:

a. Establish and strengthen agricultural land-use and land-resource planning, management, education and information at national and local levels;

b. Initiate and maintain district and village agricultural land-resource planning, management and conservation groups to assist in problem identification, development of technical and management solutions, and project implementation.

(b) Data and Information

38. Governments at the appropriate level, with the support of the relevant international and regional organizations, should:

a. Collect, continuously monitor, update and disseminate information, whenever possible, on the utilization of natural resources and living conditions, climate, water and soil factors, and on land use, distribution of vegetation cover and animal species, utilization of wild plants, production systems and yields, costs and prices, and social and cultural considerations that affect agricultural and adjacent land use;

b. Establish programmes to provide information, promote discussion and encourage the formation of management groups.

(c) International and Regional Cooperation and Coordination

39. The appropriate United Nations agencies and regional organizations should:

a. Strengthen or establish international, regional and subregional technical working groups with specific terms of reference and budgets to promote the integrated use of land resources for agriculture, planning, data collection and diffusion of simulation models of production and information dissemination;

b. Develop internationally acceptable methodologies for the establishment of databases, description of land uses and multiple goal optimization.

Means of implementation

(a) Financing and Cost Evaluation

40. The Conference secretariat has estimated the average total annual cost (1993-2000) of implementing the activities of this programme to be about $1.7 billion, including about $250 million from the international community on grant or concessional terms. These are indicative and order-of-magnitude estimates only and have not been reviewed by Governments. Actual costs and financial terms, including any that are non-concessional, will depend upon, inter alia, the specific strategies and programmes Governments decide upon for implementation.

(b) Scientific and Technological Means

41. Governments at the appropriate level, with the support of the relevant international and regional organizations, should:

a. Develop databases and geographical information systems to store and display physical, social and economic information pertaining to agriculture, and the definition of ecological zones and development areas;

b. Select combinations of land uses and production systems appropriate to land units through multiple goal optimization procedures, and strengthen delivery systems and local community participation;

c. Encourage integrated planning at the watershed and landscape level to reduce soil loss and protect surface and groundwater resources from chemical pollution.

(c) Human Resource Development

42. Governments at the appropriate level, with the support of the relevant international and regional organizations, should:

a. Train professionals and planning groups at national, district and village levels through formal and informal instructional courses, travel and interaction;

b. Generate discussion at all levels on policy, development and environmental issues related to agricultural land use and management, through media programmes, conferences and seminars.

(d) Capacity-building

43. Governments at the appropriate level, with the support of the relevant international and-regional organizations, should:

a. Establish land-resource mapping and planning units at national, district and village levels to act as focal points and links between institutions and disciplines, and between Governments and people;

b. Establish or strengthen Governments and international institutions with responsibility for agricultural resource survey, management and development; rationalize and strengthen legal frameworks; and provide equipment and technical assistance.

E. Land Conservation and Rehabilitation

Basis for Action

44. Land degradation is the most important environmental problem affecting extensive areas of land in both developed and developing countries. The problem of soil erosion is particularly acute in developing countries, while problems of salinization, waterlogging, soil pollution and loss of soil fertility are increasing in all countries. Land degradation is serious because the productivity of huge areas of land is declining just when populations are increasing rapidly and the demand on the land is growing to produce more food, fibre and fuel. Efforts to control land degradation, particularly in developing countries, have had limited success to date. Well planned, long-term national and regional land conservation and rehabilitation programmes, with strong

political support and adequate funding, are now needed. While land-use planning and land zoning, combined with better land management, should provide long-term solutions, it is urgent to arrest land degradation and launch conservation and rehabilitation programmes in the most critically affected and vulnerable areas.

Objectives

45. The objectives of this programme area are:

a. By the year 2000, to review and initiate, as appropriate, national land-resource surveys, detailing the location, extent and severity of land degradation;

b. To prepare and implement comprehensive policies and programmes leading to the reclamation of degraded lands and the conservation of areas at risk, as well as improve the general planning, management and utilization of land resources and preserve soil fertility for sustainable agricultural development.

Activities

(a) Management-related Activities

46. Governments at the appropriate level, with the support of the relevant international and regional organizations, should:

a. Develop and implement programmes to remove and resolve the physical, social and economic causes of land degradation, such as land tenure, appropriate trading systems and agricultural pricing structures, which lead to inappropriate land-use management;

b. Provide incentives and, where appropriate and possible, resources for the participation of local communities in the planning, implementation and maintenance of their own conservation and reclamation programmes;

c. Develop and implement programmes for the rehabilitation of land degraded by water-logging and salinity;

d. Develop and implement programmes for the progressive use of non-cultivated land with agricultural potential in a sustainable way.

(b) Data and Information

47. Governments, at the appropriate level, with the support of the relevant international and regional organizations, should:

a. Conduct periodic surveys to assess the extent and state of its land resources;

b. Strengthen and establish national land-resource data banks, including identification of the location, extent and severity of existing land degradation, as well as areas at risk, and evaluate the progress of the conservation and rehabilitation programmes launched in this regard;

c. Collect and record information on indigenous conservation and rehabilitation practices and farming systems as a basis for research and extension programmes.

(c) International and Regional Cooperation and Coordination

48. The appropriate United Nations agencies, regional organizations and non-governmental organizations should:

a. Develop priority conservation and rehabilitation programmes with advisory services to Governments and regional organizations;

b. Establish regional and subregional networks for scientists and technicians to exchange experiences, develop joint programmes and spread successful technologies on land conservation and rehabilitation.

Means of Implementation

(a) Financing and Cost Evaluation

49. The Conference secretariat has estimated the average total annual cost (1993-2000) of implementing the activities of this programme to be about $5 billion, including about $800 million from the international community on grant or concessional terms. These are indicative and order-of-magnitude estimates only and have not been reviewed by Governments. Actual costs and financial terms, including any that are non-concessional, will depend upon, inter alia, the specific strategies and programmes Governments decide upon for implementation.

(b) Scientific and Technological Means

50. Governments at the appropriate level, with the support of the relevant international and regional organizations, should help farming household communities to investigate and promote site-specific technologies and farming systems that conserve and rehabilitate land, while increasing agricultural production, including conservation tillage agroforestry, terracing and mixed cropping.

(c) Human Resource Development

51. Governments at the appropriate level, with the support of the relevant international and regional organizations, should train field staff and land users in indigenous and modern techniques of conservation and rehabilitation and should establish training facilities for extension staff and land users.

(d) Capacity-building

52. Governments at the appropriate level, with the support of the relevant international and regional organizations, should:

a. Develop and strengthen national research institutional capacity to identify and implement effective conservation and rehabilitation practices that are appropriate to the existing socio-economic physical conditions of the land users;
b. Coordinate all land conservation and rehabilitation policies, strategies and programmes with related ongoing programmes, such as national environment action plans, the Tropical Forestry Action Plan and national development programmes.

F. Water for Sustainable Food Production and Sustainable Rural Development

53. This programme area is included in chapter 18 (Protection of the quality and supply of freshwater resources), programme area F.

G. Conservation and sustainable utilization of plant genetic resources for food and sustainable agriculture

Basis for Action

54. Plant genetic resources for agriculture (PGRFA) are an essential resource to meet future needs for food. Threats to the security of these resources are growing, and efforts to conserve, develop and use genetic diversity are underfunded and understaffed. Many existing gene banks provide inadequate security and, in some instances, the loss of plant genetic diversity in gene banks is as great as it is in the field.

55. The primary objective is to safeguard the world's genetic resources while preserving them to use sustainably. This includes the development of measures to facilitate the conservation and use of plant genetic resources, networks of in situ conservation areas and use of tools such as ex situ collections and germ plasma banks. Special emphasis could be placed on the building of endogenous capacity for characterization, evaluation and utilization of PGRFA, particularly for the minor crops and other underutilized or non-utilized species of food and agriculture, including tree species for agro-forestry. Subsequent action could be aimed at consolidation and efficient management of networks of in situ conservation areas and use of tools such as ex situ collections and germ plasma banks.

56. Major gaps and weaknesses exist in the capacity of existing national and international mechanisms to assess, study, monitor and use plant genetic resources to increase food production. Existing institutional capacity, structures and programmes

are generally inadequate and largely underfunded. There is genetic erosion of invaluable crop species. Existing diversity in crop species is not used to the extent possible for increased food production in a sustainable way.4

Objectives

57. The objectives of this programme area are:

a. To complete the first regeneration and safe duplication of existing ex situ collections on a world-wide basis as soon as possible;

b. To collect and study plants useful for increasing food production through joint activities, including training, within the framework of networks of collaborating institutions;

c. Not later than the year 2000, to adopt policies and strengthen or establish programmes for in situ on-farm and ex situ conservation and sustainable use of plant genetic resources for food and agriculture, integrated into strategies and programmes for sustainable agriculture;

d. To take appropriate measures for the fair and equitable sharing of benefits and results of research and development in plant breeding between the sources and users of plant genetic resources.

Activities

(a) Management-related Activities

58. Governments at the appropriate level, with the support of the relevant international and regional organizations, should:

a. Develop and strengthen institutional capacity, structures and programmes for conservation and use of PGRFA;

b. Strengthen and establish research in the public domain on PGRFA evaluation and utilization, with the objectives of sustainable agriculture and rural development in view;

c. Develop multiplication/propagation, exchange and dissemination facilities for PGRFAs (seeds and planting materials), particularly in developing countries and monitor, control and evaluate plant introductions;

d. Prepare plans or programmes of priority action on conservation and sustainable use of PGRFA, based, as appropriate, on country studies on PGRFA;

e. Promote crop diversification in agricultural systems where appropriate, including new plants with potential value as food crops;

f. Promote utilization as well as research on poorly known, but potentially useful, plants and crops, where appropriate;

g. Strengthen national capabilities for utilization of PGRFA, plant breeding and seed production capabilities, both by specialized institutions and farming communities.

(b) Data and Information

59. Governments at the appropriate level, with the support of the relevant international and regional organizations, should:

a. Develop strategies for networks of in situ conservation areas and use of tools such as on-farm ex situ collections, germplasm banks and related technologies;
b. Establish ex situ base collection networks;
c. Review periodically and report on the situation on PGRFA, using existing systems and procedures;
d. Characterize and evaluate PGRFA material collected, disseminate information to facilitate the use of PGRFA collections and assess genetic variation in collections.

(c) International and Regional Cooperation and Coordination

60. The appropriate United Nations agencies and regional organizations should:

a. Strengthen the Global System on the Conservation and Sustainable Use of PGRFA by, inter alia, accelerating the development of the Global Information and Early Warning System to facilitate the exchange of information; developing ways to promote the transfer of environmentally sound technologies, in particular to developing countries; and taking further steps to realize farmers' rights;
b. Develop subregional, regional and global networks of PGRFA in situ in protected areas;
c. Prepare periodic state of the world reports on PGRFA;
d. Prepare a rolling global cooperative plan of action on PGRFA;
e. Promote, for 1994, the Fourth International Technical Conference on the Conservation and Sustainable Use of PGRFA, which is to adopt the first state of the world report and the first global plan of action on the conservation and sustainable use of PGRFA;
f. Adjust the Global System for the Conservation and Sustainable Use of PGRFA in line with the outcome of the negotiations of a convention on biological diversity.

Means of Implementation

(a) Financing and Cost Evaluation

61. The Conference secretariat has estimated the average total annual cost (1993-2000) of implementing the activities of this programme to be about $600 million, including about $300 million from the international community on grant or concessional terms. These are indicative and order-of-magnitude estimates only and have not been reviewed by Governments. Actual costs and financial terms, including any that are non-concessional, will depend upon, inter alia, the specific strategies and programmes Governments decide upon for implementation.

(b) Scientific and Technological Means

62. Governments, at the appropriate level, with the support of the relevant international and regional organizations, should:

a. Develop basic science research in such areas as plant taxonomy and phytogeography, utilizing recent developments, such as computer sciences, molecular genetics and in vitro cryopreservation;

b. Develop major collaborative projects between research programmes in developed and developing countries, particularly for the enhancement of poorly known or neglected crops;

c. Promote cost-effective technologies for keeping duplicate sets of ex situ collections (which can also be used by local communities);

d. Develop further conservation sciences in relation to in situ conservation and technical means to link it with ex situ conservation efforts.

(c) Human Resource Development

63. Governments at the appropriate level and with the support of the relevant international and regional organizations should:

a. Promote training programmes at both undergraduate and post-graduate levels in conservation sciences for running PGRFA facilities and for the design and implementation of national programmes in PGRFA;

b. Raise the awareness of agricultural extension services in order to link PGRFA activities with user communities;

c. Develop training materials to promote conservation and utilization of PGRFA at the local level.

(d) Capacity-building

64. Governments at the appropriate level, with the support of the relevant

international and regional organizations, should establish national policies to provide legal status for and strengthen legal aspects of PGRFA, including long-term financial commitments for germplasm collections and implementation of activities in PGRFA.

H. Conservation and Sustainable Utilization of Animal Genetic Resources for Sustainable Agriculture

Basis for Action

65. The need for increased quantity and quality of animal products and for draught animals calls for conservation of the existing diversity of animal breeds to meet future requirements, including those for use in biotechnology. Some local animal breeds, in addition to their socio-cultural value, have unique attributes for adaptation, disease resistance and specific uses and should be preserved. These local breeds are threatened by extinction as a result of the introduction of exotic breeds and of changes in livestock production systems.

Objectives

66. The objectives of this programme area are:

a. To enumerate and describe all breeds of livestock used in animal agriculture in as broad a way as possible and begin a 10-year programme of action;

b. To establish and implement action programmes to identify breeds at risk, together with the nature of the risk and appropriate preservation measures;

c. To establish and implement development programmes for indigenous breeds in order to guarantee their survival, avoiding the risk of their being replaced by breed substitution or cross-breeding programmes.

Activities

(a) Management-related Activities

67. Governments at the appropriate level, with the support of the relevant international and regional organizations, should:

a. Draw up breed preservation plans, for endangered populations, including semen/embryo collection and storage, farm-based conservation of indigenous stock or in situ preservation;

b. Plan and initiate breed development strategies;

c. Select indigenous populations on the basis of regional importance and genetic uniqueness, for a 10-year programme, followed by selection of an additional cohort of indigenous breeds for development.

(b) Data and Information

68. Governments at the appropriate level, with the support of the relevant international and regional organizations, should prepare and complete national inventories of available animal genetic resources. Cryogenic storage could be given priority over characterization and evaluation. Training of nationals in conservation and assessment techniques would be given special attention.

(c) International and regional cooperation and coordination

69. The appropriate United Nations and other international and regional agencies should:

a. Promote the establishment of regional gene banks to the extent that they are justified, based on principles of technical cooperation among developing countries;

b. Process, store and analyse animal genetic data at the global level, including the establishment of a world watch list and an early warning system for endangered breeds; global assessment of scientific and intergovernmental guidance of the programme and review of regional and national activities; development of methodologies, norms and standards (including international agreements); monitoring of their implementation; and related technical and financial assistance;

c. Prepare and publish a comprehensive database of animal genetic resources, describing each breed, its derivation, its relationship with other breeds, effective population size and a concise set of biological and production characteristics;

d. Prepare and publish a world watch list on farm animal species at risk to enable national Governments to take action to preserve endangered breeds and to seek technical assistance, where necessary.

Means of Implementation

(a) Financing and Cost Evaluation

70. The Conference secretariat has estimated the average total annual cost (1993-2000) of implementing the activities of this programme to be about $200 million, including about $100 million from the international community on grant or concessional terms. These are indicative and order-of-magnitude estimates only and have not been reviewed by Governments. Actual costs and financial terms, including any that are non-concessional, will depend upon, inter alia, the specific strategies and programmes Governments decide upon for implementation.

(b) Scientific and Technological Means

71. Governments at the appropriate level, with the support of the relevant international and regional organizations, should:

a. Use computer-based data banks and questionnaires to prepare a global inventory/world watch list;

b. Using cryogenic storage of germplasm, preserve breeds at serious risk and other material from which genes can be reconstructed.

(c) Human Resource Development

72. Governments at the appropriate level, with the support of the relevant international and regional organizations, should:

a. Sponsor training courses for nationals to obtain the necessary expertise for data collection and handling and for the sampling of genetic material;

b. Enable scientists and managers to establish an information base for indigenous livestock breeds and promote programmes to develop and conserve essential livestock genetic material.

(d) Capacity-building

73. Governments at the appropriate level, with the support of the relevant international and regional organizations, should:

a. Establish in-country facilities for artificial insemination centres and in situ breeding farms;

b. Promote in-country programmes and related physical infrastructure for animal livestock conservation and breed development, as well as for strengthening national capacities to take preventive action when breeds are endangered.

I. Integrated pest management and control in agriculture

Basis for Action

74. World food demand projections indicate an increase of 50 per cent by the year 2000 which will more than double again by 2050. Conservative estimates put pre-harvest and post-harvest losses caused by pests between 25 and 50 per cent. Pests affecting animal health also cause heavy losses and in many areas prevent livestock development. Chemical control of agricultural pests has dominated the scene, but its overuse has adverse effects on farm budgets, human health and the environment, as well as on international trade. New pest problems continue to develop. Integrated pest management, which combines biological control, host plant resistance and appropriate farming practices and minimizes the use of pesticides, is the best option for the future,

as it guarantees yields, reduces costs, is environmentally friendly and contributes to the sustainability of agriculture. Integrated pest management should go hand in hand with appropriate pesticide management to allow for pesticide regulation and control, including trade, and for the safe handling and disposal of pesticides, particularly those that are toxic and persistent.

Objectives

75. The objectives of this programme area are:

a. Not later than the year 2000, to improve and implement plant protection and animal health services, including mechanisms to control the distribution and use of pesticides, and to implement the International Code of Conduct on the Distribution and Use of Pesticides;

b. To improve and implement programmes to put integrated pest-management practices within the reach of farmers through farmer networks, extension services and research institutions;

c. Not later than the year 1998, to establish operational and interactive networks among farmers, researchers and extension services to promote and develop integrated pest management.

Activities

(a) Management-related Activities

76. Governments at the appropriate level, with the support of the relevant international and regional organizations, should:

a. Review and reform national policies and the mechanisms that would ensure the safe and appropriate use of pesticides — for example, pesticide pricing, pest control brigades, price-structure of inputs and outputs and integrated pest-management policies and action plans;

b. Develop and adopt efficient management systems to control and monitor the incidence of pests and disease in agriculture and the distribution and use of pesticides at the country level;

c. Encourage research and development into pesticides that are target-specific and readily degrade into harmless constituent parts after use;

d. Ensure that pesticide labels provide farmers with understandable information about safe handling, application and disposal.

(b) Data and Information

77. Governments at the appropriate level, with the support of the relevant international and regional organizations, should:

a. Consolidate and harmonize existing information and programmes on the use of pesticides that have been banned or severely restricted in different countries;

b. Consolidate, document and disseminate information on biological control agents and organic pesticides, as well as on traditional and other relevant knowledge and skills regarding alternative non-chemical ways of controlling pests;

c. Undertake national surveys to establish baseline information on the use of pesticides in each country and the side-effects on human health and environment, and also undertake appropriate education.

(c) International and Regional Cooperation and Coordination

78. Appropriate United Nations agencies and regional organizations should:

a. Establish a system for collecting, analysing and disseminating data on the quantity and quality of pesticides used every year and their impact on human health and the environment;

b. Strengthen regional interdisciplinary projects and establish integrated pest management (IPM) networks to demonstrate the social, economic and environmental benefits of IPM for food and cash crops in agriculture;

c. Develop proper IPM, comprising the selection of the variety of biological, physical and cultural controls, as well as chemical controls, taking into account specific regional conditions.

Means of Implementation

(a) Financing and Cost Evaluation

79. The Conference secretariat has estimated the average total annual cost (1993-2000) of implementing the activities of this programme to be about $1.9 billion, including about $285 million from the international community on grant or concessional terms. These are indicative and order-of-magnitude estimates only and have not been reviewed by Governments. Actual costs and financial terms, including any that are non-concessional, will depend upon, inter alia, the specific strategies and programmes Governments decide upon for implementation.

(b) Scientific and Technological Means

80. Governments at the appropriate level, with the support of the relevant international and regional organizations, should launch on-farm research in the development of non-chemical alternative pest management technologies.

(c) Human Resource Development

81. Governments at the appropriate level, with the support of the relevant international and regional organizations, should:

a. Prepare and conduct training programmes on approaches and techniques for integrated pest management and control of pesticide use, to inform policy makers, researchers, non-governmental organizations and farmers;

b. Train extension agents and involve farmers and women's groups in crop health and alternative non-chemical ways of controlling pests in agriculture.

(d) Capacity-building

82. Governments at the appropriate level, with the support of the relevant international and regional organizations, should strengthen national public administrations and regulatory bodies in the control of pesticides and the transfer of technology for integrated pest management.

J. Sustainable plant nutrition to increase food production

Basis for Action

83. Plant nutrient depletion is a serious problem resulting in loss of soil fertility, particularly in developing countries. To maintain soil productivity, the FAO sustainable plant nutrition programmes could be helpful. In sub-Saharan Africa, nutrient output from all sources currently exceeds inputs by a factor of three or four, the net loss being estimated at some 10 million metric tons per year. As a result, more marginal lands and fragile natural ecosystems are put under agricultural use, thus creating further land degradation and other environmental problems. The integrated plant nutrition approach aims at ensuring a sustainable supply of plant nutrients to increase future yields without harming the environment and soil productivity.

84. In many developing countries, population growth rates exceed 3 per cent a year, and national agricultural production has fallen behind food demand. In these countries the goal should be to increase agricultural production by at least 4 per cent a year, without destroying the soil fertility. This will require increasing agricultural production in high-potential areas through efficiency in the use of inputs. Trained labour, energy supply, adapted tools and technologies, plant nutrients and soil enrichment will all be essential.

Objectives

85. The objectives of this programme area are:

a. Not later than the year 2000, to develop and maintain in all countries the integrated plant nutrition approach, and to optimize availability of fertilizer and other plant nutrient sources;

b. Not later than the year 2000, to establish and maintain institutional and human infrastructure to enhance effective decision-making on soil productivity;

c. To develop and make available national and international know-how to farmers, extension agents, planners and policy makers on environmentally sound new and existing technologies and soil-fertility management strategies for application in promoting sustainable agriculture.

Activities

(a) Management-related Activities

86. Governments at the appropriate level, with the support of the relevant international and regional organizations, should:

a. Formulate and apply strategies that will enhance soil fertility maintenance to meet sustainable agricultural production and adjust the relevant agricultural policy instruments accordingly;

b. Integrate organic and inorganic sources of plant nutrients in a system to sustain soil fertility and determine mineral fertilizer needs;

c. Determine plant nutrient requirements and supply strategies and optimize the use of both organic and inorganic sources, as appropriate, to increase farming efficiency and production;

d. Develop and encourage processes for the recycling of organic and inorganic waste into the soil structure, without harming the environment, plant growth and human health.

(b) Data and Information

87. Governments at the appropriate level, with the support of the relevant international and regional organizations, should:

a. Assess "national accounts" for plant nutrients, including supplies (inputs) and losses (outputs) and prepare balance sheets and projections by cropping systems;

b. Review technical and economic potentials of plant nutrient sources, including national deposits, improved organic supplies, recycling, wastes, topsoil produced from discarded organic matter and biological nitrogen fixation.

(c) International and Regional Cooperation and Coordination

88. The appropriate United Nations agencies, such as FAO, the international agricultural research institutes, and non-governmental organizations should collaborate in carrying out information and publicity campaigns about the integrated plant nutrients approach, efficiency of soil productivity and their relationship to the environment.

Means of Implementation

(a) Financing and Cost Evaluation

89. The Conference secretariat has estimated the average total annual cost (1993-2000) of implementing the activities of this programme to be about $3.2 billion, including about $475 million from the international community on grant or concessional terms. These are indicative and order-of-magnitude estimates only and have not been reviewed by Governments. Actual costs and financial terms, including any that are non-concessional, will depend upon, inter alia, the specific strategies and programmes Governments decide upon for implementation.

(b) Scientific and Technological Means

90. Governments at the appropriate level, with the support of the relevant international and regional organizations, should:

a. Develop site-specific technologies at benchmark sites and farmers' fields that fit prevailing socio-economic and ecological conditions through research that involves the full collaboration of local populations;

b. Reinforce interdisciplinary international research and transfer of technology in cropping and farming systems research, improved in situ biomass production techniques, organic residue management and agroforestry technologies.

(c) Human Resource Development

91. Governments at the appropriate level, with the support of the relevant international and regional organizations, should:

a. Train extension officers and researchers in plant nutrient management, cropping systems and farming systems, and in economic evaluation of plant nutrient impact;

b. Train farmers and women's groups in plant nutrition management, with special emphasis on topsoil conservation and production.

(d) Capacity-building

92. Governments at the appropriate level, with the support of the relevant international and regional organizations, should:

a. Develop suitable institutional mechanisms for policy formulation to monitor and guide the implementation of integrated plant nutrition programmes through an interactive process involving farmers, research, extension services and other sectors of society;

b. Where appropriate, strengthen existing advisory services and train staff, develop and test new technologies and facilitate the adoption of practices to upgrade and maintain full productivity of the land.

K. Rural Energy Transition to Enhance Productivity

Basis for Action

93. Energy supplies in many countries are not commensurate with their development needs and are highly priced and unstable. In rural areas of the developing countries, the chief sources of energy are fuelwood, crop residues and manure, together with animal and human energy. More intensive energy inputs are required for increased productivity of human labour and for income-generation. To this end, rural energy policies and technologies should promote a mix of cost-effective fossil and renewable energy sources that is itself sustainable and ensures sustainable agricultural development. Rural areas provide energy supplies in the form of wood. The full potential of agriculture and agroforestry, as well as common property resources, as sources of renewable energy, is far from being realized. The attainment of sustainable rural development is intimately linked with energy demand and supply patterns.5

Objectives

94. The objectives of this programme area are:

a. Not later than the year 2000, to initiate and encourage a process of environmentally sound energy transition in rural communities, from unsustainable energy sources, to structured and diversified energy sources by making available alternative new and renewable sources of energy;

b. To increase the energy inputs available for rural household and agro-industrial needs through planning and appropriate technology transfer and development;

c. To implement self-reliant rural programmes favouring sustainable development of renewable energy sources and improved energy efficiency.

Activities

(a) Management-related Activities

95. Governments at the appropriate level, with the support of the relevant international and regional organizations, should:

a. Promote pilot plans and projects consisting of electrical, mechanical and thermal power (gasifiers, biomass, solar driers, wind-pumps and combustion systems) that are appropriate and likely to be adequately maintained;

b. Initiate and promote rural energy programmes supported by technical training, banking and related infrastructure;

c. Intensify research and the development, diversification and conservation of energy, taking into account the need for efficient use and environmentally sound technology.

(b) Data and Information

96. Governments at the appropriate level, with the support of the relevant international and regional organizations, should:

a. Collect and disseminate data on rural energy supply and demand patterns related to energy needs for households, agriculture and agro-industry;

b. Analyse sectoral energy and production data in order to identify rural energy requirements.

(c) International and Regional Cooperation and Coordination

97. The appropriate United Nations agencies and regional organizations should, drawing on the experience and available information of non-governmental organizations in this field, exchange country and regional experience on rural energy planning methodologies in order to promote efficient planning and select cost-effective technologies.

Means of Implementation

(a) Financing and Cost Evaluation

98. The Conference secretariat has estimated the average total annual cost (1993-2000) of implementing the activities of this programme to be about $1.8 billion per year, including about $265 million from the international community on grant or concessional terms. These are indicative and order-of-magnitude estimates only and have not been reviewed by Governments. Actual costs and financial terms, including any that are non-concessional, will depend upon, inter alia, the specific strategies and programmes Governments decide upon for implementation.

(b) Scientific and Technological Means

99. Governments at the appropriate level, with the support of the relevant international and regional organizations, should:

a. Intensify public and private sector research in developing and industrialized countries on renewable sources of energy for agriculture;

b. Undertake research and transfer of energy technologies in biomass and solar energy to agricultural production and post-harvest activities.

(c) Human Resource Development

100. Governments at the appropriate level, with the support of the relevant international and regional organizations, should enhance public awareness of rural

energy problems, stressing the economic and environmental advantages of renewable energy sources.

(d) Capacity-building

101. Governments at the appropriate level, with the support of the relevant international and regional organizations, should:

a. Establish national institutional mechanisms for rural energy planning and management that would improve efficiency in agricultural productivity and reach the village and household level;

b. Strengthen extension services and local organizations to implement plans and programmes for new and renewable sources of energy at the village level.

L. Evaluation of the Effects of Ultraviolet Radiation on Plants and animals Caused by the Depletion of the Stratospheric Ozone Layer

Basis for Action

102. The increase of ultraviolet radiation as a consequence of the depletion of the stratospheric ozone layer is a phenomenon that has been recorded in different regions of the world, particularly in the southern hemisphere. Consequently, it is important to evaluate its effects on plant and animal life, as well as on sustainable agricultural development.

Objective

103. The objective of this programme area is to undertake research to determine the effects of increased ultraviolet radiation resulting from stratospheric ozone layer depletion on the Earth's surface, and on plant and animal life in affected regions, as well as its impact on agriculture, and to develop, as appropriate, strategies aimed at mitigating its adverse effects.

Activities

Management-related Activities

104. In affected regions, Governments at the appropriate level, with the support of the relevant international and regional organizations, should take the necessary measures, through institutional cooperation, to facilitate the implementation of research and evaluation regarding the effects of enhanced ultraviolet radiation on plant and animal life, as well as on agricultural activities, and consider taking appropriate remedial measures.

AGENDA 21 (CHAPTER 32): STRENGTHENING THE ROLE OF FARMERS PROGRAMME AREA

Basis for Action

32.1 Agriculture occupies one third of the land surface of the Earth, and is the central activity for much of the world's population. Rural activities take place in close contact with nature, adding value to it by producing renewable resources, while at the same time becoming vulnerable to overexploitation and improper management.

32.2 The rural household, indigenous people and their communities, and the family farmer, a substantial number of whom are women, have been the stewards of much of the Earth's resources. Farmers must conserve their physical environment as they depend on it for their sustenance. Over the past 20 years there has been impressive increase in aggregate agricultural production. Yet, in some regions, this increase has been outstripped by population growth or international debt or falling commodity prices. Further, the natural resources that sustain farming activity need proper care, and there is a growing concern about the sustainability of agricultural production systems.

32.3 A farmer-centred approach is the key to the attainment of sustainability in both developed and developing countries and many of the programme areas in Agenda 21 address this objective. A significant number of the rural population in developing countries depend primarily upon small-scale, subsistence-oriented agriculture based on family labour. However, they have limited access to resources, technology, alternative livelihood and means of production. As a result, they are engaged in the overexploitation of natural resources, including marginal lands.

32.4 The sustainable development of people in marginal and fragile ecosystems is also addressed in Agenda 21. The key to the successful implementation of these programmes lies in the motivation and attitudes of individual farmers and government policies that would provide incentives to farmers to manage their natural resources efficiently and in a sustainable way. Farmers, particularly women, face a high degree of economic, legal and institutional uncertainties when investing in their land and other resources. The decentralization of decision-making towards local and community organizations is the key in changing people's behaviour and implementing sustainable farming strategies. This programme area deals with activities which can contribute to this end.

Objectives

32.5 The following objectives are proposed:

(a) To encourage a decentralized decision-making process through the creation and strengthening of local and village organizations that would delegate power and responsibility to primary users of natural resources;

(b) To support and enhance the legal capacity of women and vulnerable groups with regard to access, use and tenure of land;

(c) To promote and encourage sustainable farming practices and technologies;

(d) To introduce or strengthen policies that would encourage self-sufficiency in low-input and low-energy technologies, including indigenous practices, and pricing mechanisms that internalize environmental costs;

(e) To develop a policy framework that provides incentives and motivation among farmers for sustainable and efficient farming practices;

(f) To enhance the participation of farmers, men and women, in the design and implementation of policies directed towards these ends, through their representative organizations.

Activities

(a) Management-related activities

32.6 National Governments should:

(a) Ensure the implementation of the programmes on sustainable livelihoods, agriculture and rural development, managing fragile ecosystems, water use in agriculture, and integrated management of natural resources;

(b) Promote pricing mechanisms, trade policies, fiscal incentives and other policy instruments that positively affect individual farmer's decisions about an efficient and sustainable use of natural resources, and take full account of the impact of these decisions on household food security, farm incomes, employment and the environment;

(c) Involve farmers and their representative organizations in the formulation of policy;

(d) Protect, recognize and formalize women's access to tenure and use of land, as well as rights to land, access to credit, technology, inputs and training;

(e) Support the formation of farmers' organizations by providing adequate legal and social conditions.

32.7 Support for farmers' organizations could be arranged as follows:

(a) National and international research centres should cooperate with farmers' organizations in developing location-specific environment-friendly farming techniques;

(b) National Governments, multilateral and bilateral development agencies and non-governmental organizations should collaborate with farmers' organizations in formulating agricultural development projects to specific agro-ecological zones.

(b) Data and information

32.8 Governments and farmers' organizations should:

(a) Initiate mechanisms to document, synthesize and disseminate local knowledge, practices and project experiences so that they will make use of the lessons of the past when formulating and implementing policies affecting farming, forest and fishing populations;

(b) Establish networks for the exchange of experiences with regard to farming that help to conserve land, water and forest resources, minimize the use of chemicals and reduce or reutilize farm wastes;

(c) Develop pilot projects and extension services that would seek to build on the needs and knowledge base of women farmers.

(c) International and regional cooperation

32.9 FAO, IFAD, WFP, the World Bank, the regional development banks and other international organizations involved in rural development should involve farmers and their representatives in their deliberations, as appropriate.

32.10 Representative organizations of farmers should establish programmes for the development and support of farmers' organizations, particularly in developing countries.

Means of implementation

(a) Financing and cost evaluation

32.11 The financing needed for this programme area is estimated in chapter 14 (Promoting sustainable agriculture and rural development), particularly in the programme area entitled "Ensuring people's participation and promoting human resource development for sustainable agriculture". The costs shown under chapters 3 (Combating poverty), 12 (Managing fragile ecosystems: combating desertification and drought), and 13 (Managing fragile ecosystems: sustainable mountain development) are also relevant to this programme area.

(b) Scientific and technological means

32.12 Governments and appropriate international organizations, in collaboration with national research organizations and non-governmental organizations should, as appropriate:

(a) Develop environmentally sound farming technologies that enhance crop yields, maintain land quality, recycle nutrients, conserve water and energy and control pests and weeds;

(b) Conduct studies of high-resource and low-resource agriculture to compare their productivity and sustainability. The research should preferably be conducted under various environmental and sociological settings;

(c) Support research on mechanization that would optimize human labour and animal power and hand-held and animal-drawn equipment that can be easily operated and maintained. The development of farm technologies should take into account farmers' available resources and the role of animals in farming households and the ecology.

(c) Human resource development

32.13 Governments, with the support of multilateral and bilateral development agencies and scientific organizations, should develop curricula for agricultural colleges and training institutions that would integrate ecology into agricultural science. Interdisciplinary programmes in agricultural ecology are essential to the training of a new generation of agricultural scientists and field-level extension agents.

(d) Capacity-building

32.14 Governments should, in the light of each country's specific situation:

(a) Create the institutional and legal mechanisms to ensure effective land tenure to farmers. The absence of legislation indicating land rights has been an obstacle in taking action against land degradation in many farming communities in developing countries;

(b) Strengthen rural institutions that would enhance sustainability through locally managed credit systems and technical assistance, local production and distribution facilities for inputs, appropriate equipment and small-scale processing units, and marketing and distribution systems;

(c) Establish mechanisms to increase access of farmers, in particular women and farmers from indigenous groups, to agricultural training, credit and use of improved technology for ensuring food security.

The Future of Agriculture Depends on Biodiversity

On World Food Day 2004, FAO stresses that safeguarding and using the potential and diversity of nature is critical for world food security Scientists have identified about 1.4 million unique species of plants and animals on the planet — so far. Nearly every day, it seems, a new species is added to the list.

This variety of life is essential to human existence. We depend on it for food, water, energy, shelter, and in numerous other ways as well. But as the planet's human population continues to expand, this biodiversity is coming under increasing threat.

The most visible harm is caused by damage to natural habitats. Wild species become extinct when the places where they live are destroyed. Pollution, urbanization, deforestation and conversion of wetlands force out wildlife. Mismanagement of agriculture, forestry and fisheries can further accelerate this destructive process.

At the same time, over 40 percent of Earth's surface is comprised of agricultural lands — biodiversity is just as important on farms and in fields as it is in deep river valleys or mountain cloud forests.

This *agricultural biological diversity* consists of countless farmed plants that feed and heal people, crop varieties and aquatic species with specific nutritional characteristics, livestock species adapted to harsh environments, insects that pollinate fields and micro-organisms that regenerate agricultural soils.

But on the farm, too, biodiversity is at risk. Humankind increasingly depends on a reduced amount of agricultural biological diversity for its food supplies.

Worrisome Trends

Some 10 000 years ago, humanity began a great experiment. Using the natural biodiversity around them, they started harvesting wild seeds and plants and domesticating them, choosing those varieties that yielded the most food, or the best rope, or which did well even in drought years.

Around the same time, they began domesticating animals as well, harnessing their power, eating their meat and drinking their milk.

The diversity of the plants and animals the first farmers had to work with allowed them to select strains of plants and breeds of animals specifically tailored to meet specific needs.

Today, genetic diversity remains essential for the continued sustainability of world agricultural production.

Farmers and agricultural scientists need it in order to adapt plants to changing conditions or expand production in new, previously unfarmed areas. The hidden genetic diversity of plants holds the key to improved yields, and crops that not only produce more to eat, but more nutritious food as well.

Yet currently, four plant species — wheat, maize, rice and potato — provide over half of plant-based calories in the human diet, while around a dozen species of animals provide 90% of the animal protein consumed globally.

Beyond the number of species used to produce food — known as *species diversity* — *genetic diversity* within species and populations is crucial as well. Demand from a growing, urbanized population, however, has encouraged many farmers to adopt uniform high-yielding types of plants or animals. When food producers abandon diversity, varieties and breeds may die out — along with specialized traits.

Diversity Brings Benefits

This rapidly diminishing gene pool worries experts. Having a broad range of unique characteristics allows plants and animals to be bred to meet changing conditions. It also gives scientists the raw materials they need to develop more productive and resilient crop varieties and breeds.

In those places where hunger is worst, the resource-poor countries of the developing world, farmers may be more likely to need crops that grow well in harsh climates, rather than strains that yield well under good conditions, or animals that are smaller but possess higher resistance to disease. Indeed, for the poorest farmers, the diversity of life may be their best protection against starvation.

Consumers in the developed and developing world alike also benefit when they have access to a wide choice of plants and animals. This contributes to a nutritious diet, particularly important for rural communities with limited access to markets.

Finally, with plants, animals and their environments left intact, a range of essential services provided by nature are preserved. Livestock, fungi and micro-organisms decompose organic matter, transferring nutrients to the soil. Ants and other insects control pest populations. Bees, butterflies, birds and bats pollinate fruit trees. Swamps and marshes filter out pollutants. Forests prevent flooding and reduce erosion. In the ocean, intact ecosystems help keep fish populations stable and healthy, ensuring tomorrow's catch.

Guaranteeing the future

To feed a growing population, agriculture must provide more food. It will also be essential to increase its resilience by protecting a wide array of life forms with unique traits, such as plants that survive drought or livestock that reproduce in harsh conditions. Sustainable agricultural practices can both feed people and protect the oceans, forests, prairies and other ecosystems that harbour biological diversity.

Global efforts to conserve plants and animals in gene banks, botanical gardens and zoos are vital. But an equally important task is to maintain biodiversity on farms and in nature, where it can evolve and adapt to changing conditions or competition with other species. As custodians of the world's biodiversity, farmers can develop and maintain local plants and trees and reproduce indigenous animals, ensuring their survival.

To learn more about agricultural biodiversity and how FAO is working with countries around the world to preserve it, read the related articles that make up this Focus on the issues package, linked to the right.

Biodiversity Action Plan

A Biodiversity Action Plan (BAP) is an internationally recognized programme addressing threatened species and habitats, which is designed to protect and restore biological systems. The original impetus for these plans derives from the 1992 Convention on Biological Diversity (CBD). As of 2006, 188 countries have ratified the CBD, but only a fraction of these have developed substantive BAP documents.

The principal elements of a BAP typically include: (a) preparing inventories of biological information for selected species or habitats; (b) assessing the conservation status of species within specified ecosystems; (c) creation of targets for conservation and restoration; and (d) establishing budgets, timelines and institutional partnerships for implementing the BAP.

Species Plans

A fundamental method of engagement to a BAP is thorough documentation regarding individual species, with emphasis upon the population distribution and conservation status. This task, while fundamental, is highly daunting, since only an estimated ten percent of the world's species are believed to have been characterized as of 2006, most of these unknowns being plants and lower animals. For many mammal, bird and reptile species, information is often available in published literature; however, for many plant species as well as invertebrate species, such information may require considerable local data collection. It is also useful to compile time trends of population estimates in order to understand the dynamics of population variability and vulnerability. In some parts of the world complete species inventories are not realistic; for example, in the Madagascar dry deciduous forests, many species are completely undocumented and much of the region has never even been systematically explored by scientists.

A species plan component of a country's BAP should ideally entail a thorough description of the range, habitat, behaviour, breeding and interaction with other species. Once a determination has been made of conservation status (e.g. rare, endangered, threatened, vulnerable), a plan can then be created to conserve and restore the species population to target levels. Examples of programmatic protection elements are: habitat restoration; protection of habitat from urban development; establishment of property ownership; limitations on grazing or other agricultural encroachment into habitat; reduction of slash and burn agricultural practises; outlawing killing or collecting the species; restrictions on pesticide use; and control of other environmental pollution. The plan should also articulate which public and private agencies should implement the protection strategy and indicate budgets available to execute this strategy.

Habitat Plans

Where a number of threatened species depend upon a specific habitat, it may be appropriate to prepare a habitat protection element of the Biodiversity Action Plan. Examples of such special habitats are: raised acidic bogs of Scotland; Waterberg Biosphere bushveld in South Africa; California's coastal wetlands; and Sweden's Stora Alvaret on the island of Öland. In this case also, careful inventories of species and also the geographic extent and quality of the habitat must be documented. Then, as with species plans, a program can be created to protect, enhance and/or restore habitat using similar strategies as discussed above under the species plans.

Specific Countries

Some examples of individual countries which have produced substantive Biodiversity Action Plans follow. It is worth noting that preparation of a country BAP may cost up to 100 million pounds sterling, with annual maintenance costs roughly ten percent of the initial cost. Obviously costs for countries with small geographical area or simplified ecosystems have a much lesser cost. For example the St. Lucia BAP has been costed in the area of several million pounds sterling.

Australia

Australia has developed a detailed and rigourous Biodiversity Action Plan, having identified a total of over 475,000 indigenous species. A key element of the BAP is protection of the Great Barrier Reef, which is actually in a much higher state of health than most of the world's reefs, Australia having one of the highest percentages of treated wastewater.

Considerable analysis has been conducted on the sustainable yield of firewood production, a major threat to deforestation in most tropical countries. Biological inventory work; assessment of harvesting practices; and computer modeling of the dynamics of treefall, rot and harvest; have been carried out to adduce data on safe harvesting rates. Extensive research has also been conducted on the relation of brush clearance to biodiversity decline and impact on water tables; for example, these effects have been analyzed in the Lake Toolibin wetlands region.

St. Lucia

The St. Lucia BAP recognizes impacts of large numbers of tourists to the marine and coastal diversity of the Soufrière area of the country. The BAP specifically acknowledges that the carrying capacity for human use and water pollution discharge of sensitive reef areas was exceeded by the year 1990. The plan also addresses conservation of the historic island fishing industry. In 1992, several institutions in conjunction with native fishermen to produce a sustainable management plan for fishery resources, embodied in the Soufrière Marine Management Area.

The St. Lucia BAP features significant involvement from the University of the West Indies. Specific detailed attention is given to three species of threatened marine turtles, to a variety of vulnerable birds and a number of pelagic fishes and cetaceans. In terms of habitat conservation the plan focusses attention on the biologically productive mangrove swamps and notes that virtually all mangrove areas had already come under national protection by 1984.

Tanzania

The Tanzania national BAP addresses issues related to sustainable use of Lake Manyara, an extensive freshwater lake, whose usage by humans accelerated in the period 1950 to 1990. The designation of the Lake Manyara Biosphere Reserve combines conservation of the lake and surrounding high value forests with sustainable use of the wetlands area and simple agriculture. This BAP has united principal lake users in establishing management targets. The Biosphere Reserve has induced sustainable management of the wetlands, including monitoring groundwater and the chemistry of the escarpment water source.

United Kingdom

The United Kingdom Biodiversity Action Plan covers not only terrestrial species associated with lands within the UK, but also marine species and migratory birds, which spend a limited time in the UK or its offshore waters. The UK plan encompasses "391 Species Action Plans, 45 Habitat Action Plans and 162 Local Biodiversity Action Plans with targeted actions". This plan is noteworthy because of its extensive detail, clarity of endangerment mechanisms, specificity of actions, follow up monitoring program and its inclusion of migrating cetaceans and pelagic birds.

United States of America

Twenty-six years prior to the international biodiversity convention, the USA had launched a national program to protect threatened species in the form of the 1966 Endangered Species Act. The legislation created broad authority for analyzing and listing species of concern, and mandated that Species Recovery Plans be created. Thus, while the USA is an unratified signer of the accord, arguably it has the longest track record and most comprehensive program of species protection of any country. There are approximately 7000 listed species (e.g. endangered or threatened), of which about half have approved Recovery Plans. While this number of species seems high compared to other countries, the value is rather indicative of the total number of species characterized, which is extremely large.

Uzbekistan

Five major divisions of habitat have been identified in Uzbekistan's BAP: Wetlands (including reed habitat and man-made marsh); desert ecosystems (including sandy, stony and clay); steppes; riparian ecosystems; and mountain ecosystems. Over 27,000 species have been inventoried in the country, with a high rate of endemism for fishes and reptiles. Principal threats to biodiversity are related to human activities associated with overpopulation and generally related to agricultural intensification. Major geographic regions encompassed by the BAP include the Aral Sea Programme (threatened by long-term drainage and salination, largely for cotton production), the Nuratau Biosphere Reserve, and the Western Tien Shan Mountains Programme (in conjunction with Kazakhstan and Kyrgyzstan).

Criticism of Biodiversity Action Plans

Some developing countries criticize the emphasis of BAPs, because these plans inherently favour consideration of wildlife and plant protection above food and industrial production, and in some cases may represent an obstacle to population growth. The plans are costly to produce, a fact which makes it difficult for many smaller countries and poorer countries to comply. In terms of the plans themselves, many countries have adopted pro-forma plans including little research and even less in the way of natural resource management. With regard to specific world regions, there is a notable lack of substantive participation by most of the Middle Eastern countries and much of Africa, the latter of which may be impeded by economic considerations of plan preparation. Some governments such as the European Union have diverted the purpose of a Biodiversity Action Plan, and implemented the convention accord by a set of economic development policies with referencing certain ecosystems' protection.

Common Agricultural Policy

The Common Agricultural Policy (CAP) is a system of European Union agricultural subsidies and programmes. It represents about 44% of the EU's budget (•43 billion scheduled spending for 2005. These subsidies work by guaranteeing a minimum price to producers and by direct payment of a subsidy for crops planted. Reforms of the system are currently underway, including a phased transfer of subsidy to land stewardship rather than specific crop production from 2005 to 2012. Detailed implementation of the scheme varies in different member countries of the EU, but currently a new Single Payment Scheme for direct farm payments is being introduced in the UK.

Beginnings of the CAP

The creation of a common agricultural policy was proposed in 1960 by the European Commission. It followed the signing of the Treaty of Rome in 1957, which established the Common Market. The six member states individually strongly intervened in their agricultural sectors, in particular with regard to what was produced, maintaining prices for goods and how farming was organised. This intervention posed an obstacle to free trade in goods while the rules continued to differ from state to state, since freedom of trade would interfere with the intervention policies. Some Member States, in particular France, and all farming professional organizations wanted to maintain strong state intervention in agriculture. This could therefore only be achieved if policies were harmonised and transferred to the European Community level.

By 1962, three major principles had been established to guide the CAP; market unity, community preference and financial solidarity. Since then, the CAP has been a central element in the European institutional system.

The CAP is often explained as the result of a political compromise between France and Germany: German industry would have access to the French market; in exchange, Germany would help pay for France's farmers.

Objectives

Sectors Covered by the CAP

The common agricultural policy price intervention covers only certain agricultural products:

- cereal, rice, potatoes
- oilseeds
- dried fodder
- milk and milk products, wine, honey
- beef and veal, poultry meat and eggs, pig meat, sheep / lamb meat and goat meat
- sugar
- fruit and vegetables
- cotton
- peas, field beans
- sweet lupins
- olive oil
- seed flax
- silkworms
- fibre flax
- hemp
- tobacco
- hops
- seeds
- flowers and live plants
- animal feed stuffs

The coverage of products in the external trade regime is more extensive than the coverage of the CAP regime. This is to limit competition between EU products and alternative external goods (for example, litchi juice could potentially compete with orange juice).

The initial objectives were set out in Article 33 of the Treaty of Rome:

1. to increase productivity, by promoting technical progress and ensuring the optimum use of the factors of production, in particular labour;
2. to ensure a fair standard of living for the agricultural Community;
3. to stabilise markets;
4. to secure availability of supplies;
5. to provide consumers with food at reasonable prices.

The CAP recognised *the need to take account of the social structure of agriculture and of the structural and natural disparities between the various agricultural regions and to effect the appropriate adjustments by degrees.*

How the CAP Works

CAP is an integrated system of measures which works by maintaining commodity price levels within the EU and by subsidising production. There are three principal mechanisms:

- Import Tariffs are applied to specified goods imported into the EU. These are set at a level to raise the World market price up to the EU target price. The target price is chosen as the maximum desirable price for those goods within the EU.
- An internal intervention price is set. If the internal market price falls below the intervention level then the EU will buy up goods to raise the price to the intervention level. The intervention price is set lower than the target price. The internal market price can only vary in the range between the intervention price and target price.
- Subsidies are paid to farmers growing particular crops. This was intended to encourage farmers to choose to grow those crops attracting subsidies and maintain home-grown supplies. Subsidies were generally paid on the area of land growing a particular crop, rather than on the total amount of crop produced. Current reforms of the system now underway are phasing out specific crop subsidies in favour of flat-rate subsidies based only on the area of land in cultivation, and for adopting environmentally beneficial farming methods. This will reduce, but not eliminate, the economic incentive to overproduce.

The change in subsidies is intended to be accomplished by 2011, but individual governments have freedom to decide how the new scheme will be introduced. The UK government has decided to run both systems of subsidy together, each year transferring a larger proportion of the total payments to the new scheme. Other governments have chosen to wait, and change the system in one go at the latest possible time.

The CAP also makes use of external trade policy. Some non member countries have negotiated quotas which allow them to sell particular goods within the EU without tariffs. This notably applies to countries which had a traditional trade link with a member country.

The CAP also aims to promote legislative harmonisation within the Community. Differing laws in member countries can create problems for anyone seeking to trade between countries. Examples are regulations on permitted preservatives or coloring agents in foods, labelling regulations, use of hormones or other drugs in livestock intended for human consumption and disease control (e.g. during the foot and mouth disease outbreak in the United Kingdom, Ireland and the Netherlands), animal welfare regulations. The process of removing all hidden legislative barriers to trade is still incomplete. The CAP is funded by the European Agricultural Guidance and Guarantee Fund (EAGGF) of the EU. CAP reform has steadily lowered its share of the EU budget but it still accounts for nearly half EU expenditure. In recent years France has benefited the most from these subsidies. The new accession countries which joined the EU in 2004 have large farm sectors and would have overtaken France as chief beneficiary, but for transitional regulations limiting the subsidies which they receive. The continuing problem of how subsidies for these countries will be paid when they become eligible has already led to French concessions on reform of the CAP. Further concessions will inevitably be necessary to balance the budget.

History of the CAP

The European Community (predecessor of the EU) rose from the ashes of World War II. It was set up to end the old hostilities that led to wars and to create prosperity through cooperation among all Europeans. The Common Agricultural Policy (CAP) was born in the late 1950s and early 1960s when the founding members of the EU had just emerged from over a decade of severe food shortages during and after the Second World War. As part of building a common market, tariffs on agriculture would have to be removed. However, due to the political clout of farmers, and the sensitivity of the issue, it would take many years before the CAP was fully implemented.

The Treaty of Rome defined the general objectives of a common agricultural policy. The principles of the Common Agricultural Policy (CAP) were set out at the Stresa Conference in July 1958. In 1960, the CAP mechanisms were adopted by the six founding Member States and two years later, in 1962, the CAP came into force.

By 1962, three major principles had been established to guide the CAP; market unity, community preference and financial solidarity. Since then, the CAP has been an important element in the European institutional system.

Evolution and Reform

The CAP has always been a difficult area of EU policy to reform; this is a problem that began in the 1960s and one that continues to the present day, albeit less severely. The Agricultural Council is the main decision-making body for CAP affairs and is dextrously manipulated by those states that hold the CAP most dearly, such as France. Above all, however, unanimity is needed for most serious CAP reform votes, resulting in rare and gradual change. Outside Brussels proper, the farming lobby's power has been a factor determining EU agricultural policy since the earliest days of integration. This lobby's power has decreased markedly since the 1980s, but even today some attempts at reform are prevented by this group.

In recent times change has been more forthcoming, due to external trade demands and intrusion in CAP affairs by other parts of the EU policy framework, such as consumer advocate working groups and the environmental departments of the Union. In addition Euroscepticism in states such as the UK and Denmark is fed in part by the CAP, which is actually detrimental to their economies.

Keeping the CAP intact, though, is an important aim of EU policy. Farming is regarded as "special", a part of Europe's shared heritage encompassing food production and even fine dining. All of these are used as rationales for keeping the CAP strong. It is not simply just another industry, hence its massive presence in the EU psyche (and the EU budget.) Finally, the aim of self-sufficiency and a "shared larder" in Europe, a particularly salient concern in the post-war years, lingers to this day.

Pre-1992

Early attempts at reforms, such as the Mansholt Plan, tended to fail. The Mansholt Plan was a 1960s idea that sought to remove small farmers from the land and to consolidate farming into a larger, more efficient industry. Farming's special status, and above all the extremely powerful farming lobbies across the Continent saw the Plan disappear from the Union's objectives.

The Mansholt Plan: On 21 December 1968, Sicco Mansholt, European Commissioner for Agriculture, sent a memorandum to the Council of Ministers concerning agricultural reform in the European Community. This long-term plan, also known as the '1980 Agricultural Programme' or the 'Report of the Gaichel Group', named after the village in Luxembourg where it had been quietly prepared, laid the foundations for a new social and structural policy for European agriculture.

The Mansholt Plan noted the limits to a policy of price and market support. It predicted the imbalance that would occur in certain markets unless the Community undertook to reduce its land under cultivation by at least 5 million hectares. The former Netherlands Minister of Agriculture also noted that the standard of living of farmers had not improved since the implementation of the CAP, despite an increase in production and permanent increases in Community expenditure. He therefore suggested that production methods should be reformed and modernised and that small farms, which were bound to disappear sooner or later, according to Community experts, should be increased in size. The aim of the Plan was to encourage nearly five million farmers to give up farming. That would make it possible to redistribute their land and increase the size of the remaining family farms. Farms were considered viable if they could guarantee for their owners an average annual income comparable to that of all the other workers in the region. In addition to vocational training measures, Mansholt also provided for welfare programmes to cover retraining and early retirement. Finally, he called on the Member States to limit direct aid to unprofitable farms.

Faced with the increasingly angry reaction of the agricultural community, Sicco Mansholt was soon forced to reduce the scope of some of his proposals. Ultimately, the Mansholt Plan was reduced to just three European directives which, in 1972, concerned the modernisation of agricultural holdings, the abandonment of farming and the training of farmers. Bruised by the failure of Mansholt, would-be reformers were mostly absent throughout the 1970s, and reform proposals were few and far between if at all. Not least due to the various financial crises that rocked the union in this decade, such as oil supply problems and the economic depression. A system called "Agrimoney" was introduced as part of the fledgling EMU project, but was deemed a failure and did not stimulate further reforms.

The 1980s was the decade that saw the first true reforms of the CAP, foreshadowing further development from 1992 onwards. The influence of the farming bloc declined, and with it, reformers were emboldened. Environmentalists garnered great support in reining in the CAP, but it was financial matters that ultimately tipped the balance: due to huge overproduction the CAP was becoming expensive and wasteful. These factors combined saw the introduction of a quota on dairy production in 1984, and finally, in 1988, a ceiling on EU expenditure to farmers. However, the basis of the CAP remained in place, and not until 1992 did CAP reformers begin to work in earnest.

1992

In 1992, the MacSharry reforms (named after the European Commissioner for Agriculture, Ray MacSharry) were created to limit rising production, while at the same time adjusting to the trend toward a more free agricultural market. The reforms reduced levels of support by 29% for cereals and 15% for beef. They also created 'set-aside' payments to withdraw land from production, payments to limit stocking levels, and introduced measures to encourage retirement and forestation.

Since the MacSharry reforms cereal prices are closer to the equilibrium level, there is greater transparency in costs of agricultural support and the 'de-coupling' of income support from production support has begun. However, the administrative complexity involved invites fraud, and the associated problems of the CAP are far from being corrected.

It is worth noting that one of the main catalysts behind the 1992 reforms was the need to pacify the EU's external trade partners at the Uruguay round of the GATT trade talks with regards to agricultural subsidies. This set the tone for later reforms which were more often than not direct responses to external pressures on the Union, as opposed to a genuine and spirited response to the various anti-CAP groups existing within the EU.

2003

On 26 June 2003, EU farm ministers adopted a fundamental reform of the CAP, based on almost entirely "decoupling" subsidies from a particular crop. (Though Member States may choose to maintain a limited amount of specific subsidy.) The new "single farm payments" are linked to respect for environmental, food safety and animal welfare standards. The aim is to make more money available for environmental, quality or animal welfare programmes by reducing direct payments for bigger farms.

Details of the UK scheme were still being decided at its introductory date of May 2005. Details of the scheme in each member country may be varied subject to outlines issued by the EU. In the UK the single payment scheme provides a single flat rate payment of around £230 per hectare for maintaining land in cultivateable condition. This will be phased in from 2005 to 2012 such that each year an increasing proportion of subsidy is paid under the new scheme. The remaining proportion will be paid under the pre-2005 scheme which provided different subsidies for different crops. The new scheme allows for much wider non-production use of land which may still receive subsidy. Additional payments are available if land is managed in ecologically friendly ways.

The overall EU and national budgets for subsidy have been capped. This will prevent growth in the total bill to the taxpayer.

The reforms enter into force in 2004-2005. (Member States may apply for a transitional period delaying the reform in their country to 2007 and phasing in reforms up to 2012)

EU Expansion 2004

The expansion of the EU in 2004 increased the number of farmers from 7 to 11 million, increased the agricultural land area by 30% and crop production by 10-20%. The 2004 entrants into the EU have immediate access to price support measures (export refunds, intervention buying). However direct payments will be phased in over

10 years (2004-2013), starting at 25% of the rate paid to existing countries in 2004, and 30% for 2005. The 2004 entrants to the EU have access to a rural development fund (for early retirement, environmental issues, poorest areas, technical assistance) with a •5 billion budget. EU states agreed in 2002 that agricultural expenditure up to 2013 should not increase in real terms. This will require a cut in subsidies to the original states of around 5% to finance payments to the new members. With Romania and Bulgaria joining in 2007, the required cut will increase to 8%.

The current areas that are issues of reform in EU agriculture are: lowering prices, ensuring food safety and quality, and guaranteeing stability of farmers' incomes. Other issues are environmental pollution, animal welfare, and finding alternative income opportunities for farmers. Some of these issues are the responsibility of the member states.

European Commission Report

A 2003 report, commissioned by the European Commission, by a group of experts led by Belgian economist André Sapir stated that the budget structure was a "historical relic". The report suggested a rethink of EU policy, redirecting expenditure towards measures intended to increase wealth creation and cohesion of the EU. As a significant proportion of the budget is currently spent on agriculture, and there is little prospect of the budget being increased, this would necessitate reducing CAP expenditure. The report largely concerned itself discussing alternative measures more useful to the EU, rather than discussing the CAP, but it did also suggest that farm aid would be administered more effectively by member countries on an individual basis.

The report's findings were largely ignored. Instead, CAP spending was kept within the remit of the EU — and France led an effort to agree a fixed arrangement for CAP spending that would not be changed until 2012. This was made possible by advance agreement to this approach with Germany. It is this agreement that the UK currently wishes to see re-opened, both in their efforts to defend the UK position on the UK rebate and also given that the UK is in favour of lowering barriers to entry for third world agricultural exporters.

Sugar Regime Reform 2005

One of the crops subsidised by the CAP is sugar produced from sugar beet; the EU is by far the largest sugar beet producer in the world, with annual production at 17 million metric tons. This compares to levels produced by Brazil and India, the two largest producers of sugar from sugar cane.

Sugar was not included in the 1992 McSherry reform, or in the 1999 Agenda 2000 decisions; sugar was also subject to a phase-in (to 2009) under the Everything But Arms trade deal giving market access to least developed countries. In 2005 European Union agriculture ministers announced plans to cut the minimum beet price by 39%

from 2006, over four years. Under the Sugar Protocol to the Lome Convention, nineteen ACP countries export sugar to the EU, and will be affected by price reductions on the EU market.

These proposals followed the WTO appellate body largely upholding on 28 April 2005 the initial decision against the EU sugar regime.

2006 Reforms

As of 21 February 2006, the EU has decided on some reforms of sugar subsidies. The guaranteed price of sugar is to be cut by 36%, with European production projected to fall sharply as a result of this. According to the EU, this is the first serious reform of sugar under the CAP for 40 years.

An aim of this policy change is to allow easier and more profitable access to European markets for emerging economies. Critics, such as "EUPolitix", contend that this is not an altruistic move nor an idealistic shift from the EU, who are instead acting only in accordance with the wishes of the WTO, who supported challenges on sugar dumping by the EU from Australia, Thailand and Brazil. Another point of contention is that those countries who currently receive preferential treatment from EU member states — often due to colonial ties — as part of the ACP group may stand to lose out.

Criticism of the CAP

The CAP has been roundly criticised by many diverse interests since its inception. Criticism has been wide-ranging, and the even the European Commission has long been persuaded of the numerous defects of the policy.

Anti-development

Criticism of the CAP has united some supporters of globalisation with the anti-globalisation movement in that it is argued that these subsidies, like those of the USA and other Western states, add to the problem of what is sometimes called Fortress Europe; the West spends high amounts on agricultural subsidies every year, which amounts to unfair competition. The OECD countries' total agricultural subsidies amount to more than the GDP of the whole of Africa.

Moreover, it is argued that in creating an oversupply of agricultural products which are then sold in the Third World and simultaneously preventing the Third World from exporting its agricultural goods to the West, the CAP increases Third World poverty by putting Third World farmers out of business. According to the *Human Development Report 2003* in 2000 the average dairy cow in the EU received \$913 in subsidies, compared with an average of \$8 per person in Sub-Saharan Africa.

Artificially High Food Prices

CAP price intervention causes artificially high food prices throughout the EU. Some have suggested that Europeans pay about 25% higher prices for food than they would without the CAP, whereas the Timbro research institute has counted figures reaching over 80%. Others have calculated the rate at somewhere around 15%, and argue that most calculations are inflated because they usually compare actual consumer prices in Europe with "ideal" market prices in the rest of the world, not actual consumer prices. Some commodities have even more inflated prices: European sugar costs more than three times the global market price. This subsidy is estimated to cost each EU citizen on average £16 or •24 per week although intervention costs and subsidy are decreasing.

As a consequence of these artificial high food prices European consumers pay a regressive consumption tax. In general, the fraction of income spent on food products tends to increase as income decreases, as everyone has to eat at least a minimum of food. Because the relative spending on food is higher among the lower income classes, they pay relatively more taxes than higher income classes. (Baldwin & Wyplosz, *The Economics of European Integration*, McGraw-Hill Education, Maidenhead (Berkshire), 2004, pp. 218-219.)

Hurting Smaller Farms

Although most policy makers in Europe agree that they want to promote "family farms" and smaller scale production, the CAP in fact rewards larger producers. Because the CAP has traditionally rewarded farmers who produce more, larger farms have benefitted much more from subsidies than smaller farms. For example, a farm with 1000 hectares, earning one hundred extra euros per hectare will make 100,000 extra euros, while a 10 hectare farm will only make an extra 1000 euros, disregarding economies of scale. As a result most CAP subsidies have made their way to large scale farmers. Since the 2003 reforms subsidies have been linked to the size of farms, so this is not so true any more. So while subsidies allow small farms to exist, they funnel most profits to larger scale operations.

Environmental Problems

The CAP has traditionally promoted a large expansion in agricultural production. At the same time it has allowed farmers to employ unecological ways of increasing production, such as the indiscriminate use of fertilizers and pesticides, with serious environmental consequences. Since the reforms of 2003 much more attention is paid to environmental problems, and this may no longer be true.

Equity Among Member States

France has a slightly lower GDP than the UK, and its higher population means that it earns slightly less per person compared to the UK. Germany has a GDP approximately 25% higher than either France or the UK, but per capita income is comparable to the other two countries. France makes a net payment into the EU budget, so it can not be said that it receives a subsidy from any other country. Rather, France, Germany and the UK all contribute towards funding of CAP subsidies to other member states, France contributing the least of the three.

UK Rebate and the CAP

The UK would have been contributing considerably more money to the EU budget, except that Margaret Thatcher's government successfully negotiated a special annual UK rebate in 1984. Without the rebate the UK was the largest contributor despite being the third poorest member state. Due to the way the rebate is funded, France pays the largest share of the rebate (31%), followed by Italy (24%) and Spain (14%).

The discrepency in CAP funding is a cause of some consternation in the UK. As of 2004, France received 13% of total CAP funds more than the UK. This is a net benefit to France of •6.37 billion, compared to the UK. This is largely a reflection of the fact that France has more than double the land area of the UK. In comparison, the UK budget rebate for 2005 is scheduled to be approx •5.5 billion. The popular view in the UK (as, for example, set forth in the tabloid press) is that if the UK rebate were reduced with no change to the CAP, then the UK would be paying money to keep an inefficient French farming sector in business — to many British people, this would be seen as grossly unfair. French motives for generating arguments about "solidarity" and "selfishness" are therefore seen as extremely self-serving.

If the rebate were removed without changes to the CAP then the UK would pay a net contribution of 14 times that of the French (In 2005 EU budget terms). The UK would make a net contribution of •8.25 billion compared to the current contribution of •2.75 billion, versus a current French net contribution of •0.59 billion.

In December 2005 the UK agreed to give up approximately 20% of the rebate for the period 2007-2013, on condition that the funds did not contribute to CAP payments, were matched by other countries contributions and were only for the new member states. Spending on the CAP remained fixed, as had previously been agreed. Overall, this reduced the proportion of the budget spent on the CAP. It was agreed that the European Commission should conduct a full review of all EU spending.

CAP Financing set up to Benefit France?

Many critics allege that the CAP has been of special benefit to France, which has a large agricultural sector. It is widely acknowledged that the CAP was initially established largely to suit France's interests, as it hoped to expand its large agricultural sector in which it enjoyed comparative advantage. However, many other countries have benefited from CAP subsidies, especially those with large agricultural sectors. Also, over the years France has made many concessions on various issues. The CAP is certainly not exclusively designed to benefit France, and farmer's associations across the union have supported it. In fact the German government's reliance on agricultural swing votes has been vital to maintaining CAP programming through the decades.

In their book, *The great deception: can the European Union survive?*, Christopher Booker and Richard North argue that the fundamental reason that France's President De Gaulle kept Britain out of the EEC during the 1960s was his concern to have the financial arrangement for the Common Agricultural Policy established first.

According to Booker and North, De Gaulle manipulated the EEC to get them to underwrite the high subsidies for French farmers, who in 1961 still accounted for a quarter of France's employment as opposed to only four percent in Britain. Once the CAP funding was settled, British membership of the EEC became a matter of French interest, and De Gaulle's veto was abandoned. As a condition of its membership, Britain cut her imports of cheap food from around the world and replaced them with more expensive French and continental products. At the same time the levies that it paid on what foodstuffs it imported from outside the EEC were automatically transferred to Brussels to subsidise French and other EEC farmers.

State Intervention

Some major critics of the Common Agricultural Policy reject the idea of protectionism, either in theory, practice or both. Free market advocates are among those who disagree with government intervention because, they say, a free market without interference will allocate resources more efficiently. The major intervention during the 1980's was the creation of Grain Mountains, where huge stores of grain were bought directly from farmers at a price set by the government of the day. Subsidies allow many small, outdated, or inefficient farms to continue to operate which would not otherwise be viable. A straightforward economic model would suggest that it would be better to allow the market to find its own price levels, and for uneconomic farming to cease. Resources used in farming would then be switched to more productive operations. This is important.

Economic Sustainability

Many economists believe that the CAP is unsustainable in an enlarged EU. The inclusion of ten additional countries on May 1, 2004 has obliged the EU to take measure to limit CAP expenditure. Poland is the largest new member and has two million smallhold farmers. It is significantly larger than any of the other new members, but taken together the new states represent a significant increase in recipients under the CAP. Even before expansion, the CAP consumed a very large proportion of the EU's budget, upward of 90% in the late 1980s. Considering that a small proportion of the population, and relatively small proportion of the GDP comes from farms, many considered this expense excessive.

Policy Recommendations for Sustainable Agriculture

The Task Force reached consensus on nine recommendations. These recommendations are designed to achieve the goals that the Task Force outlined for a sustainable U.S. agriculture. Thus the first five recommendations are aimed at achieving environmentally sound and economically viable agricultural production; the sixth recommendation at revitalizing rural farming communities; the seventh recommendation at producing a safe and high-quality food supply; the eighth recommendation at encouraging research on integrating productivity, profitability, and environmental stewardship into the U.S. agricultural system; and the ninth recommendation at achieving international harmonization of intellectual property rights. The recommendations do not address all aspects of the stated goals; they are, however, bold first steps in accomplishing these goals.

Each recommendation is followed by a brief discussion of the concerns that inspired it and by yardsticks for measuring progress in meeting a particular goal. These indicators are not intended to be mandates for specific actions or policies.

Goal

Management of agricultural activities to protect air, soil, and water quality, and to conserve wildlife habitat and biodiversity, thereby increasing agriculture's long-term productivity and profitability, as well as enhancing human health and well-being.

Policy Recommendation 1

Integrate pollution prevention and natural resource conservation into agricultural production

Harmonize the pollution prevention and natural resource conservation policies of various federal, state, and local agencies to minimize conflicts among the policies that could undermine environmental protection. Link technical and financial assistance to farmers and ranchers to their voluntary implementation of farm- and ranch-specific plans for integrating pollution prevention and natural resource conservation into agricultural production. Renew and refine land retirement program to improve their cost-effectiveness, increase their conservation of natural resources, and enhance their ability to prevent agriculturally related pollution.

Discussion

Of the 1.9 billion acres of U.S. land, excluding Alaska, approximately 907 million acres are croplands, pastures, or rangelands. The management of these agricultural lands can affect the quality of ecosystems and the condition of natural resources over a significant portion of the U.S. land base. For this reason, farmers and ranchers should be encouraged to control sediment carried in runoff from farms and ranches, use environmentally sound pest and nutrient management techniques, reduce consumption of nonrenewable energy, and take other actions that would preserve the health of ecosystems and conserve natural resources.

Because some federal, state, and local policies and programs relating to agriculture conflict with one another, they may actually undermine efforts to attain national and local environmental objectives and sustainable development goals, as well as inefficiently use public resources. To eliminate this conflict, all levels of government should review these policies and programs and work together to coordinate them. In doing so, they should invite nongovernmental organizations, such as university research institutions and agribusinesses, to help identify policies that potentially hinder farmers and ranchers from protecting natural resources and preventing pollution. Successful collaboration will depend on the forging of strong partnerships among the U.S. Department of Agriculture (USDA), the U.S. Department of Interior, the U.S. Environmental Protection Agency, the U.S. Axmy Corps of Engineers, state agriculture and environmental protection agencies, farm groups, universities, agribusinesses, producers' organizations, and other entities.

The USDA, in particular, should take several actions to promote environmentally sound agricultural practices. In addition to assessing whether its policies encourage sustainable agriculture and revising or eliminating those policies that do not, it should direct additional technical and financial resources to meeting natural resource and environmental protection goals, strengthen its soil and wetlands conservation efforts, and link participation in its farm programs to farmers' voluntary implementation of integrated farming systems (that is, systems that integrate pollution prevention and natural resource conservation into agricultural production). USDA also should develop projects to demonstrate integrated farming systems on both small- and large-scale farms and direct its research efforts toward promoting environmental protection and conservation of natural resources.

The government has an important role to play in resource conservation efforts relating to land retirement programs, through which agricultural lands are managed for envirorunental benefits by contracts or purchase of easements—with federal, state, or local funds, or some combination of these funds. These programs generate economic and environmental benefits for farmers and society as a whole. The economic benefits include reducing crop surpluses and federal budgetary outlays for crop subsidies. The environmental benefits—which derive from the restoration of wetlands and grasslands—include increased recreational opportunities, improved water quality and carbon sequestration, and the provision of habitat for wild plant and animal species. The Conservation Reserve Program, an existing land retirement program with enrollment at 36.4 million acres, has been credited with generating approximately $8.6 billion in wildlife-related benefits alone.

In the past, federal and state governments have designed many resource conservation programs from the top down, with inadequate local involvement. As a result, community priorities often are not heard or understood. To take these priorities into account in the implementation of conservation programs, the USDA has established state technical committees. These committees—which include federal, state, and local agency representatives—provide an opportunity for stakeholders to establish environmental protection and natural resource conservation goals and criteria for meeting them, as well as an opportunity to develop guidelines for target cost-share payments. The USDA should encourage local environmental, consumer, agricultural, and other groups to contribute input to state committees' implementation of conservation and the environmental programs.

Indicators

The Task Force identified the following indicators of progress toward integrating natural resource conservation and pollution prevention into agricultural production.

- Conflicts in government policies relating to sustainable agriculture are resolved, and policies that discourage sustainable agriculture are revised or eliminated by the year 2005.
- By the year 2000, government agencies (particularly the USDA) implement programs to encourage the voluntary implementation of plans for integrating natural resource conservation and pollution prevention into farm and ranch operations.
- The interests of a broad range of stakeholders are represented on state technical committees.
- Violations of drinking water, surface water, and soil quality standards by agricultural practices decrease.
- Land retirement programs target environmentally sensitive, marginal croplands—including riparian corridors and wetlands—so that these lands can be set aside or managed primarily to preserve their environmental values.

Policy Recommendation 2

Increase the flexibility for participants in commodity programs to respond to market signals and adopt environmentally sound production practices and systems, thereby increasing profitability and enhancing environmental quality.

Increase the flexibility for farmers who participate in commodity programs to make their own production decisions in response to market signals so that they can implement profitable production practices and systems that conserve natural resources, enhance environmental quality, and optimize resource use.

Discussio Commodity programs—through which the federal government supports prices for some crops—can distort market signals and prevent farmers from making the most efficient use of agricultural inputs and the natural resource base. As a result, the programs impose economic costs on farmers and environmental costs on the rest of the American public.

One restriction of commodities programs illustrates how these costs arise. "Base acreage" requirements, which specify the minimum number of acres that must be planted in one of the price-supported crops, encourage intensive monoculture. Therefore, the requirements discourage farmers from diversifying their crops in response to market demand and from optimizing their use of resources, which would promote environmental stewardship.

If commodity programs gave participants greater flexibility in their production decisions, farmers would be better able to manage their crops in ways that increase both profitability and environmental protection. Past experience indicates that farmers will take advantage of opportunities to do so. In 1990, Congress passed legislation that allowed farmers who had signed up for a particular commodity program—for example, the wheat program—to plant some of their land in a crop other than that specified by the program. In response, farmers reduced the number of acres under monoculture and diversified their crops. By 1994, approximately 42 percent of the land on which farmers were allowed to grow whatever they chose was planted in crops other than those specified by the commodity program in which the farmers were enrolled.

Congress should continue to enhance the flexibility of commodity programs.

Indicators

The Task Force identified the following indicators of progress toward increased flexibility in commodities programs:

- Congress continues to legislate reform of commodities programs to give farmers flexibility to manage their lands in ways that will increase both profitability and environmental protection.
- Farmers diversify production and optimize their use of resources—including land, water, fertilizers, pesticides, energy, labor, and equipment—in environmentally beneficial ways.
- Negative agricultural impacts on the environment lessen.

Policy Recommendation 3

Expand agricultural markets

Pursue efforts to expand agriculture markets in order to increase demand for agricultural products, returns farmers and ranchers, investments in environmental protection, and conservation of natural resources. Support continued negotiations on international agreements that encourage more open global markets.

Discussion

If farmers' revenues increase, they will have greater resources to invest in their farms, ranch and communities and in environmental protection and natural resource conservation. The keys to increasing revenues are controlling production costs and expanding agricultural markets.

Farmers already are highly motivated to control production costs, but they could use the assistance of the federal government in expanding agricultural markets. Working with the private sector, the federal government should explore the feasibility of revolving loans, repayable grants, or matching grants to conduct market research on, demonstrate the acceptability of, and promote the expanded use of agricultural commodities in existing domestic markets; find new uses for these commodities; and encourage the creation of businesses that utilize agricultural commodities as rawmaterials in new products.

The federal government also has a role to play in opening up global agricultural markets to U.S. farmers. As the global population grows and as demand for higher-quality diets increases in developing countries, American farmers must be in a position to compete for additional food-supply business. The government can help them by supporting continued negotiations on international agreements that discourage trade-distoring policies.

Indicators

The Task Force identified the following indicators of progress in expanding agricultural markets:

- Farmers' market share of commodities that are produced in an environmentally friendly way increases in international markets.
- New uses for agricultural commodities are found.
- Value-added, environmentally-sound agricultural industries and entrepreneurial initiatives increase.
- International treaties eliminate trade-distoring policies.

Policy Recommendation 4

Revise the pricing of public natural resources

Continue to move the pricing of public natural resources and their use toward market pricing, recognizing that there may be circumstances when investments are warranted for the public good.

Discussion

Getting the prices of resources right is a fundamental tenet of sound resource management, yet policies that result in subsidies for the use of natural resources can provide decisionmakers and resource managers with the "wrong" information regarding the value of those natural resources. Leases for the use of public natural resources—such as water, grazing lands, and forest lands—and sales of such resources, both of which sometimes amount to subsidies, illustrate one of the consequences of this wrong information. In drawing up contracts for these leases and sales and in calculating lease rates and sale prices, the government often neglects to include the full costs of making the resources available to the private sector or to consider the market prices of the resources. As a result, the government fails to recover the costs of resource use from the user.

Policies that result in subsidies for the use of natural resources can undermine sustainability because they can incorrectly signal that supplies of the resources are larger than is actually the case and because they do not reflect the rising demand for environmental values as per capita income increases. As a result, they often have supported overconsumption or low-value uses of natural resources.

To prevent natural resource policies from undermining sustainability, the government should identify and revise, where appropriate, those subsidies that encourage the use of renewable resources at a rate greater than is sustainable over the long term. In addition, in drawing up new contracts for the lease or sale of public natural resources, the government should consider the market prices of the resources and the full costs of making the resources available to the private sector. Finally, the government should recognize rising environmental values by conserving the use of public natural resources.

Indicators

The Task Force identified the following indicators of correct prices for public natural resources:

- The pricing of public natural resources moves toward market pricing.
- Long-term investments in conservation and stewardship increase.

Policy Recommendation 5

Keep prime farmlands in agricultural production

Keep prime farmlands in agricultural production by implementing (at the appropriate levels of government) rational land-use policies such as easement, zoning, taxation, financial incentive, transportation, and land development policies that reduce the encroachment of urban sprawl on prime farmlands and that otherwise seek to preserve the prime land base for U.S. agriculture.

Discussion

Highly productive or versatile farmlands are an important part of the natural resource base upon which the production of food, feed, fuel, and fiber depend. Stewardship of these prime farmlands, which include both croplands and grazing lands, is critically important to the country's economic, environmental, and social health and well-being. Yet several pressures, both internal and external to agriculture, threaten to convert prime agricultural lands to nonagricultural uses.

One of these pressures is urban sprawl. More than half of U.S. agricultural production, measured in dollars at the farm gate, comes from counties where the expanding urban fringe threatens prime farmland. Moreover, 30 percent of the nation's agricultural production comes from so-called Metropolitan Statistical Areas, where the human population exceeds 197.7 million, and 26 percent comes from adjacent counties with a population density of at least 25 people per square mile.

Another pressure derives from the fundamental disconnection between the best use of land resources and the best use of certain lands for agricultural production. As a result of this disconnection, the prime cropland is sometimes converted for housing and other uses, leaving less productive land to support future agricultural production, as well as supply certain aesthetic, cultural, and other values.

To prevent the loss of prime agricultural lands to nonagricultural uses, states and localities should make these lands the objects of farmland protection programs.

Indicators

The Task Force identified the following indicators of progress toward reserving prime farmlands for agricultural production:

- By the year 2010, states and localities identify their most strategic agricultural lands.
- By the year 2025, farmland protection programs protect these lands.

Goal

Achieve viable farmers and farm communities.

Policy Recommendation 6

Invest in rural communities' infrastructure

Help rural communities to capitalize on the economic benefits of sustainable agriculture by giving priority in rural business development and marketing programs to investments in enterprises associated with the products of sustainable agriculture. In addition, invest in rural communities' infrastructure.

Discussion

The economic viability of many rural counties and that of farms and ranches are closely tied. Mainstreet businesses in rural communities depend on the spending power of nearby farms and ranches. In turn, these operations often depend on the services of local agricultural-input suppliers and local agricultural processing, distribution, and marketing enterprises. Therefore, investments that support enterprises associated with the products of sustainable agricultural systems will help farms and ranches as well as rural communities to capitalize on the economic benefits of these systems.

Investments that directly support sustainable agriculture are not, by themselves, sufficient to curtail the exodus of residents from rural communities. To be healthy, rural communities must have at their base a solid infrastructure to support economic development. Therefore, federal and state rural business development programs and marketing programs should make investments in infrastructure that will help revitalize many rural communities—for example, investments in upgrades of bridges and roads and in modernization of medical, communication, and capital-lending systems.

Indicators

The Task Force identified the following indicators of increased investment in rural communities' infrastructure:

- The number of rural-based, value-added businesses that utilize the products of sustainable agricultural practices increases.
- Fewer unsafe bridges and roadways are found in rural areas.
- Local tax revenues and per capita incomes increase in rural areas.
- Unemployment rates decline in rural areas.

Goal

Production of a safe, high-quality, and affordable supply of food and fiber in a manner that protects and conserves natural resources.

Policy Recommendation 7

Continue Improvements in Food Safety and Quality

Assure continued improvements in the safety and quality of U.S. food products. Pursue international harmonization of food standards while maintaining the right of the United States to pursue its own high standards for food safety and quality.

Discussion

U.S. food products have long been among the safest and highest-quality food products in the world. Technologies that are being developed to reduce microbial contamination and to better monitor, sometimes on a continuous basis, the threat of microbial and bacterial contamination should enhance prospects for maintaining the safety of food products. Other technologies are improving food quality by making possible the development or design of food products that are more closely suited to the dietary needs of domestic and foreign consumers. These technologies could, in the future, lower the fat content of meats and increase the nutritional protein content of grains and the vitamin content of fruits and vegetables.

But food safety and quality does not hinge on technological advances alone; it depends significantly on government oversight of the U.S. food complex. To ensure continued improvement in food safety and quality, the USDA should assess grading and testing standards, monitoring mechanisms, and safety standards for their effectiveness in protecting public health. It also should expand programs that educate consumers about safe food-handling practices, strengthen programs that ensure the safety of food for children and other vulnerable subgroups of the population, and increase opportunities for a more varied and healthy diet — especially for low-income families. Finally, the USDA should encourage the development and use of quick field tests for food safety and quality that can be used to assess compliance with microbial contamination and chemical residue tolerances.

Efforts to enhance food safety cannot be limited to the U.S. food system because the system operates in a global market where the safety standards of some countries are lower than in the United States. Given this reality, the federal government should reinvigorate existing efforts to strengthen food safety programs in foreign food-exporting countries. In addition, as food standards worldwide are harmonized, they should be based on a scientific process that is transparent.

Goal

Creation of institutional incentives that focus public and private research, education, and technology development on integrating agricultural profitability and productivity with environmental stewardship.

Policy Recommendation 8

Promote the research needed to support a sustainable U.S. agriculture.

Develop institutional incentives and funding arrangements to promote research that shows how to integrate agricultural productivity and profitability with environmental stewardship. Where necessary, increase efforts to convey the latest research findings to local farmers and to educate them about the most promising new technologies and management strategies for achieving efficiency.

Discussion

Agricultural research has been responsible for substantial efficiency gains in agricultural production. Major advances in agricultural science and technology have helped farmers double and even triple and quadruple the per-acre yields of some crops since the mid-1930s. They also have assisted farmers in increasing the productivity of farm animals. For example, at the beginning of this century, a milk cow produced approximately 4,000 pounds of milk each year, whereas today it can produce more than 15,000 pounds over the same period.

Looking ahead to the future, the need is for research that helps farmers and ranchers to be good environmental stewards while they are increasing their productivity. In short, a sustainable U.S. agriculture will require research that focuses on integrating productiviy, profitabliliy, and environmental stewardship.

While both publicly funded and privately funded research have greatly enhanced the performance of agricultural producers in the past, two tendencies may diminish the potential contribution of research to the sustainability of U.S. agriculture in the future. First, agricultural research is often narrowly focused. Second, publicly funded research is conducted within the constraints of brief budget cycles.

Agricultural research tends to focus on individual aspects of agricultural production-for example, pest control, soil management techniques, or development of new crop varieties-rather than on whole production systems, including the ecological systems that are the settings for farming and ranching operations. Moreover, agricultural research tends to emphasize the insights of single discoplines rather than combining the expertise of multiple disciplines, including biology, chemisty, ecology, and economics. This narrow focus, which has evolved in response to institutional pressures for specialization, impedes the acquisition of knowledge that would enhance the sustainabilty of U.S. agriculture, particularly where environmental costs and values are concerned. To remedy this problem, the U.S. Department of Agriculture (in cooperation with the U.S. Environmental Protection Agency and the U.S. Department of the Interior) should take the lead in developing a plan to focus agricultural research on multiple aspects of agricultural production from the perspective of multiple disciplines. In doing so, it should seek input from universities, scientific societies, agribusinesses, grower organizations, and other organizations.

Agricultural research programs often operate on an annual budget cycle, requiring researchers to submit grant applicatins each year to continue projects. Because projects often require more than one year's work, they could lose funding before they are completed. To remedy this situation, the USDA should revise grant programs to make provisions for financial commitments of terms longer than one year. Moreover, when administrators of grant programs are deciding which projects merit such commitments, they should give priority to projects that include on-farm demonstrations of promising new tehnologies and management strategies for enhancing the sustainablity of agriculture.

Of cource, research serves little purpose unless it reaches the hands of those in a position to use it. The implication is that agricultural extension agencies should imporve, if necessary, their efforts to convey to local farmers the latest findings of sound, publicly funded agricultural research, In addition, agricultural research institutions that receive public funds should educate agricultural producers, to the extent that they are not already doing so, about the latest tehnological advances and most efficient management practices.

Indicators

The Task Force identified the following indicators of progress in generating the research needed to support a sustainable agriculture and in promoting education about promising technologies and management strategies:

- By the year 2000, USDA develops a plan to overcome institutional and funding barriers to multifaced and multidisciplinary research.
- By the year 2000, USDA develops criteria to ensure that agricultural research proposals simultaneously address issues of profitability, productivity, and environmental protection.
- By 2010, a signigicant portion and, by 2020, a majority of publicly funded research administered by USDA addresses issues of profitablilty, productivity, and environmental protection.
- All applied research has a strong technology transfer and demonstration component.
- Research grants administered by USDA are allocated on a competitive, merit basis.
- Consumers, conservation groups, and other public interests are involved in USDA's decisions about which research to fund.
- USDA substantially increases its funding of research, including on-farm studies of sustainable production systems.
- USDA expands its cooperative research with academic institutions and nongovernmental organizations on perennial polyculture systems modeled on natural ecosystems.
- Non-USDA programs that provide federal funding for agricultural research and extension programs begin assessing proposals for their relevance to sustainable agriculture.

Policy Recommendation 9

Pursue international harmonization of intellectual property rights

Pursue international harmonization of intellectual property rights in order provide incentives for the development of new agricultural technologies. Support the objectives of the International Convention on Biological Diversity in order to conserve genetic resources and protect intellectual property rights.

Discussion

Given the food and fiber demands projected for the 21st century, development of ever more advanced technologies for increading agricultural productivity is critical. But this productivity cannot jeopardize biologically diverse ecosystems. In the long term, the sustainability of U.S. agriculture depends on both new agricultural technologies and efforts to conserve biodiversity and genetic resources globally. Intellectual property rights are key to the development of these technologies and the success of these efforts.

Recent trade agreements, notably NAFTA (North American Free Trade Agreement) and those agreements reached during the Uruguay Round of GATT (General Agreement on Tariffs and Trade), have imposed new guidelines for the protection of intellectual property rights. As these guidelines are implemented and adopted by additional countries, the U.S. government should work to ensure that they maximize benefits for genetic resource conservation and sustainable agriculture. In addition, the U.S. government should help to structure and implement the intellectual property provisions of trade agreements and other international treaties in ways that protect the interests of indigenous peoples whose long-standing knowledge of biodiversity resources has helped fuel innovation and development in multibillion dollar in dustries, including agriculture, but who often have not been compensated for sharing this knowledge.

Indicators

The Task Force identified the following indicators of the influence of intellectual proprety rights:

- A plan providing incentives for the countries to adopt genetic-resource conservation provisions is developed by the end of 1996 and reflected in U.S. agricultural, trade, and foreign policies shortly thereafter.
- The United States ratifies the Convention on Biological Diversity and leads efforts for its implementation worldwide.

10

Agricultural Information Management: Case Study of Select Workshops, Conferences & Publications

Agricultural Information Management Standards

The Agricultural Information Management Standards, abbreviated to AIMS, is an initiative started by the Food and Agriculture Organization (FAO) of the United Nations. The overall aim of the initiative is to improve coherence among agricultural information systems. The immediate goal of this project is to create a clearing house for information management standards, which will help to make existing or new agricultural information systems interoperable; and allow for sharing and promoting the uptake of common methodologies, standards and applications. The AIMS website provides lists of tools to facilitate interoperability between existing agricultural datasets, and methods for creating new systems that use consistent and commonly accepted standards in the description of information related to agriculture and similar fields.

The initiative is a collaborative effort between various partner agricultural organizations which share a common goal to improve coherence amongst agricultural information systems through shared standards and tools.

Standards and Tools for Managing and Sharing Information

Metadata Standards

A metadata set consists of elements in which each element refers to a label that describes particular information about a resource. A resource here is defined as "anything that has an identity". In the context of the Document-like Information Objects, a resource could be, for example, a document, monograph, web page, project report, etc. The AgMES is an initiative that aims to encompass issues of semantic standards in the domain of agriculture with respect to description, resource discovery, interoperability and metadata exchange for different types of information resources.

Knowledge Organization Systems

Within the domain of agriculture, there are several different knowledge organizations systems (KOS) currently in use. Some of the most prominent ones are grouped and presented here, with links to their original sites. You can browse the KOS registry through:

- type of KOS
- domain area covered by the KOS

Although there exists a plethora of knowledge organization systems, there is still a need to formally structure the context and meaning of terms. An ontology takes the existing knowledge organization systems such as AGROVOC one step further by structuring the terms more formally, and by providing richer relationships between concepts than those currently provided in thesauri. The Agricultural Concept Server is the first step towards an "Ontology Service". Ontology is a new concept that is emerging from the various semantic Web initiatives, which roughly speaking can be defined as a semantic system that contains terms, the definitions of those terms, and the specification of relationships among those terms. Such a semantic system can be referred to as an "Ontology Service".

Tools

- Metadata/Cataloguing Tools: This page brings together a summary of known metadata tools used for creating metadata for various types of resources. It includes tools for entering and editing metadata, and utilities for processing, extracting, validating, and viewing metadata. Some of these tools allow for metadata to be served up in different formats, such as XML, RDF, XHTML, ISO2709 etc. Some tools are based on established metadata standards while others produce metadata based specifically on local application profiles.
- Ontology Tools: When working on an ontology project, one of the steps is to find a suitable ontology creation and maintenance tool. There are a number of ontology editors currently available to help in the process of ontology development. Some tools can help to acquire, organize, edit, visualize and maintain domain knowledge and aid in the building of an ontology. Ontologies are available in different formats: XML/RDF, OWL-Full, OWL-Lite, etc. Most tools offer the possibility to import and export ontologies in various formats, as required by the specific implementation/applications.
- Thesaurus Tools: A number of commercial thesaurus management tools are currently available on the market. However, here we have listed only the freely available tools.
- Miscellaneous Tools: This page lists tools such as validators, parsers, etc.

7TH ISSCT AGRICULTURAL MECHANIZATION WORKHSOP

Louisiana, USA 12-17 October, 2003

"Burnt vs. Green Cane Mechanization — Agricultural Engineering Challenges"

Report

Introduction

The 7th Agricultural Engineering Workshop was held from 12 to 17 October 2003 in Thibodaux, Louisiana. The Workshop was organised by the Agricultural Engineering Sectional Committee and hosted by the American Sugar Cane League (ASCL) under the auspices of the American Society of Sugar Cane Technologists (ASSCT). The workshop was attended by 54 delegates from 13 countries: Australia (3), Argentina (7), Colombia (3), England (1), Guatemala (1), Guadeloupe (1), Guyana (2), India (2), Mauritius (8), Reunion (1), South Africa (2) and Zimbabwe (1). The host country had the largest contingent with 22 delegates.

Mr Charlie Melancon, President and General Manager of the ASCL, officially welcomed the delegates at a formal reception on Sunday 12 October. The Workshop was officially opened on Monday 13 October by Mr John Gay, Chairman of the Board, ASCL, who presented an historical overview of the Louisiana sugar industry and the role that the ASCL plays in that industry.

Formal Sessions

The formal sessions were held at the Howard Johnson Lodge in Thibodaux. A total of 32 verbal and 4 poster presentations were made under the four formal session topics. Formal paper presentation sessions were held on Monday, Tuesday, Thursday and Friday. Workshop topics covered the following:

Monday, 13th Green cane versus burnt cane harvesting

Tuesday, 14th Green cane versus burnt cane harvesting continued

Thursday, 16th Land preparation, planting, crop maintenance and trash management

Friday, 17th Machinery costing and benchmarking.

Each topic was introduced by a Committee member or invited speaker, followed by presentations by the delegates, with adequate time for questions and discussions. The papers were generally of a high standard with a good mixture of practical and technical papers.

Workshop Sessions

Green Cane vs Burnt Cane Harvesting

It is estimated that less than 20% of the world's more than 1 000 millions tons of sugarcane is harvested mechanically. Harvesting systems vary widely and the choice of one system over another will depend on labour availability, labour cost, topography and climatic conditions. Many sugarcane industries are still seeking an interim step between manual and fully mechanised harvesting systems. In industries where terrain, field layout or mill receiving facilities precludes the economic use of chopper harvesters, much research is being conducted to develop suitable economic whole stalk harvesters.

It was generally agreed that, regardless of the country involved, harvesting of sugarcane without burning is difficult, as most industries expected it to be. There are still many issues that have not been resolved, particularly under certain local industry conditions. These include lower and slower harvesting rates for manual and mechanical harvesting systems respectively. Although with mechanical harvesters there is the potential for reducing harvest-to-crush delays, there is also potential for higher cane losses and poorer quality of cane being delivered to factories.

Trials have shown that by improving machine maintenance and better extractor fan speed setting an amicable balance can be found between the amount of trash delivered to the factory and acceptable cane loss. There have been some significant improvements to the gathering and feeding mechanisms and the cleaning systems of modern chopper harvesters. Tests showed that the latest chopper harvesters have increased harvesting rates, lower fuel consumption and lower extraneous matter levels in cane samples being delivered to factories.

With environmental pressure increasing to restrict burning, there is a real need to further improve machine performance, reduce the amount of extraneous matter being delivered to the factory and reduce cane losses when operating in green cane. These problems can be partly solved by improving field layouts, improving agronomic practices and developing more machine-friendly cane varieties. Another issue that raises concern is the fact that the higher the degree of mechanisation, the higher the risk of soil compaction. There is also a need to develop best management practices for individual local conditions in order to ensure economically and ergonomically sustainable cropping systems. A mechanical harvesting best practice manual has been developed in Australia and is commercially available.

The ISSCT preferred harvester testing protocol was reviewed at the workshop. It was agreed that certain sections of the protocol need clarification, while others require revision. A proposed new protocol will be drawn up and circulated to the delegates before being resubmitted to ISSCT and posting on the website.

Land Preparation, Planting, Crop Maintenance, and Trash Management

There are many factors to be considered when switching from burnt to green cane harvesting. The more important advantages of green cane harvesting are well documented. However, some of the benefits are countered by the suppressive effect of trash on ratoons, the harbouring of diseases and insects pests and the need for tillage. Addressing these issues will require inputs from agricultural engineers, agronomists and economists. Millers must also accept that with green cane harvesting fibre and other extraneous matter levels will increase, but that there are also gains to be made in the form of a fresher cane supply.

Much research has been carried out around the world to develop improved mechanical planters and no-tillage whole stalk and billet planters. Poor metering systems and poor billet quality have been identified as major problem areas that result in higher than normal seeding rates to ensure acceptable cane stands. Billet quality has been improved by research into better harvester maintenance and adjustment. The adoption of minimum and no-tillage planting systems as part of an improved cropping system is seen by many industries as a major advance.

Crop rotation is also seen as a desirable practice to improve soil heath and control pests and diseases. Work is being carried out to select the optimum crops and rotation cycles. Row width is also important to cater for irrigation and mechanisation requirements.

It was shown that improved field layout design improved mechanical harvesting machinery performances by reducing time lost by up to 30%. Furthermore, cane stool damage and soil in the cane sample were both reduced significantly. Other results showed that direct stool damage by infield machinery is reduced to a minimum when cane fields are planted at wider row spacings or alternative sets of row spacings that better accommodate infield transport track widths and assist in controlling traffic paths. Trials have shown that uncontrolled traffic systems can reduce cane yield by as much as 30 t/ha.

There are several new types of equipment available commercially for use in managing crop residues. Some of the equipment can be used to incorporate trash and tops into the interrow, while others, such as a modified road brush, can be used to remove the trash from the top of the cane rows. In areas where burning is prohibited, or where crop residues cause slow emergence or yield decline, this type of equipment provides growers with alternative management practices.

The concept of recovering most of the above-ground biomass during the harvesting operation is not new. The results of trials conducted in Florida, where combine harvester extractor fans were turned off, showed that the trash content averaged about 20% compared with 8% when extractor fans were in use. Truck payloads were reduced by as much as 38%. Harvester forward speed with the extractor fans off was 10% slower than when the fans were on, and harvester output was 13% lower when the fans were off. The lower harvester output can be ascribed to the increase in total cane and residue mass passing through the machine. Juice quality was similar for both harvesting approaches. The delegates agreed that enormous potential exists for maximising the value of total biomass of the sugarcane crop.

A mini precision farming workshop was scheduled as part of this session. Precision farming is based on the concept that variability in soil fertility, soil depth, microclimate and weed species all occur naturally and are site dependent, and all of these factors have a direct influence on crop production.

The precision farming concept is really a combination of several relatively new technologies such as mirco-computers, global positing systems (GPS), geographic information systems (GIS), information gathering hardware and software, and the automatic control of farm machinery. The system has the potential to improve yields, and low-yield areas can be better managed to increase yields. The use of variable rate technology allows for more effective and more economic nutrient use, as well as being environment friendly. Although variable rate technology is well developed, the same cannot be said for cane yield monitors on sugarcane harvesters. The use of auto steer guidance equipment was shown to be most useful in maintaining accurate steerage in green cane harvesting and avoiding cane stool damage. Delegates agreed that precision farming should be part of an overall plan to make industries more productive, competitive, sustainable and environmentally friendly in the future.

Machinery Costing and Benchmarking

It was shown that in a developing world, the use of small-scale contractors on large sugarcane estates could be cheaper and gives more satisfactory service than in-house equipment fleets or large scale contractor organisations. However, for this to take place the local infrastructure must mature and be assisted and guided by the developing industry.

The Discounted Cash Flow (DCF) method of costing agricultural machinery can be used to determine the cost of mechanised operations. The DCF method takes into account factors such as effect of inflation and interest on the time value of money, as well as the implications of various income tax considerations. The DCF method can further assist in evaluating capital equipment replacement decisions. As these factors are not fully accounted for when using the popular 'Classical' machinery costing method, the use of the DCF method results in a more accurate account of machinery costs and cash flow.

Machinery and equipment costs form a major proportion of total sugarcane production costs. It is estimated that in some industries these costs can be as high as 40%. At the previous Agricultural Engineering workshop held in Malelane, South Africa, it was agreed that a standard costing method was essential and would be beneficial when reporting machinery costs at workshops and congresses. A machinery costing and performance standards protocol, based on the 'Classical' costing method, was tabled at this workshop. The proposed protocol was accepted unanimously by the delegates, and the Chairman was asked to prepare the final draft and distribute it to the delegates prior to tabling it before the ISSCT.

Another issue that was raised at the previous workshop was that of benchmarking methods and standards. It is often extremely difficult to make meaningful cost comparisons between farming enterprises that are similar, and even more so between countries or industries that are dissimilar. It was therefore proposed that a set of standard processes and procedures be drawn up for use when carrying out a benchmarking exercise to ensure portability and effective comparisons between systems.

Workshopping Sessions

At the end of each topic's formal session, time was set aside for informal 'workshopping' sessions. The aim of the workshopping sessions was to give delegates an opportunity to raise issues not covered during the formal sessions, or to air their views on matters raised during the day's proceedings. These sessions were structured under the following headings:

- Mechanical issues
- Labour issues
- Cane quality issues
- Agronomic issues
- Best management practices
- Future research requirements.

It was generally agreed by the delegates that the workshopping sessions were useful in that they expanded on numerous issues pertaining to the day's proceedings. A summary of these sessions will be forwarded to the delegates in due course.

Business Meeting

At the business meeting held at the end of the formal sessions, membership of the Agricultural Engineering Section Committee, the venue and proposed theme for the next workshop were discussed.

Delegates wishing to be considered for membership on the committee were asked to submit their names to the Chairman. The Chairman will forward the names of potential committee members to the ISSCT Technical Program and Executive Committees for consideration. The delegates unanimously proposed that the next workshop should be hosted by Argentina. Also discussed during the business meeting was the proposed theme and topics for the Agricultural Engineering sessions at the ISSCT Congress to be held in Guatemala in 2005.

Field Tours

The purpose of the field trips was to expose delegates to the widest possible range of agronomic field practices, harvesting, infield loading and cane transport systems and mill receiving facilities in use in the Louisiana and Florida sugar industries.

On a half-day tour in Louisiana delegates visited the USDA research farm near Houma where they saw how field trials are mechanically harvested and weighed. Delegates were also shown a commercial harvesting operation being carried on a private farm where a single as well as a two row harvesters were being used. A visit to the Cameco Industries' Rebecca farm made up the last stop of the day. During the full day tour, delegates were shown a single row combine harvester modified to handle sprawled cane, various trash handling and management systems and a billet transloading operation. There was also a visit to the MA Patout Enterprise sugar factory located near Jeanerette. At the factory, delegates were shown a whole truck tippler offloading system and were given a guided tour of the only diffuser sugar extraction system presently in use in Louisiana. The tour also included a visit to the factory's cane quality testing facility.

A post-workshop one day tour of the Florida sugar industry was arranged by the Florida Sugar Cane League. During this tour mechanical harvesting, manual planting as well as three mechanical planting systems were demonstrated. Other stops included water management and water quality control, and stops to see a cane harvester fitted with auto steer and a variable rate fertiliser applicator. Finally, there was a drive-past visit to the Oakeelanta Corporation's co-generation plant, which is fuelled by baggase, wood chips and cane trash, and delegates also had a look at their sugar packaging plant.

General

The general consensus was that the Workshop was a great success, and many delegates commented on the quality of the presentations and general organisation. The workshop was well supported by the Louisiana and Florida sugar industries, and the organisers wish to thank these industries for their generous contributions.

A comprehensive report on the Workshop activities will be presented at the ISSCT Congress in Guatemala City, Guatemala, which is to be held from 30 January to 4 February 2005.

Electronic Forum on Biotechnology

FAO established the Electronic Forum on Biotechnology in Food and Agriculture in March 2000 to provide quality balanced information on agricultural biotechnology in developing countries and to make a neutral platform available for people to exchange views and experiences on this subject so that it might be possible to better understand and clarify the issues and concerns behind polarization of the debate on agricultural biotechnology for developing countries.

This publication presents a report on the first six conferences of the Forum that took place from March 2000 to May 2001. Some background to the Forum and its conferences are provided in this chapter.

Definition of Biotechnology for the Purposes of the Forum

Firstly, how is biotechnology defined for the purposes of the Forum? According to the Convention on Biological Diversity (CBD), biotechnology means "any technological application that uses biological systems, living organisms, or derivatives thereof, to make or modify products or processes for specific use". Interpreted in this broad sense, the definition covers many of the tools and techniques that are commonplace today in agriculture and food production. Interpreted in a narrow sense, as is often done and as is done in the Forum, biotechnology mainly covers technological applications involving reproductive biology or, secondly, the manipulation, or use, of the genetic material of living organisms for specific uses. This definition covers a wide range of diverse technologies including, for example, the use of molecular DNA markers, gene manipulation and gene transfer, vegetative reproduction (crops and forest trees), embryo transfer and freezing (livestock) and triploidization (fish).

Background to the Establishment of the Forum at FAO

FAO was founded in 1945 with a mandate to raise levels of nutrition and standards of living, to improve agricultural productivity, and to better the condition of rural populations. It is an intergovernmental organization with 183 member countries. One of the most important tasks that FAO carries out is to collect, analyse, interpret and disseminate relevant information. FAO serves as a clearing-house, providing farmers, scientists, government planners, traders and non-governmental organizations (NGOs) with the information they need to make rational decisions on planning, investment, marketing, research and training.

FAO should play an active part in disseminating information and promoting information exchange regarding biotechnology. It is important that member countries know which biotechnologies are available, what they can be used for, how and in which wider strategy they can be applied, and what the cost-benefit implications of using them are. The global population size has passed the six billion mark and is increasing by roughly 80 million annually. Almost all population growth is in developing countries. Since the amount of agricultural land available is limited, the increases in food production needed to feed the world's growing population must come from increasing the amount of food produced per hectare. Biotechnology, which is a collection of diverse tools that can be applied to many areas of food and agriculture, may play a role here. This collection includes scientific tools (such as genetic modification) that are sometimes considered to be highly controversial. The tools may pose ethical problems and require substantial debate among policy-makers, researchers and the public at large. Particularly in some areas of agricultural biotechnology (e.g. involving cultivation of genetically modified crops), the debate has become quite polarized and there is therefore an increasing need for quality, balanced, neutral and factual information.

To consider specifically the background to the establishment of the Forum, the biennial meeting of FAO's Committee on Agriculture (COAG), held in Rome from 25-29 January 1999, was of key importance because, among other areas, it set the direction for FAO's future involvement in biotechnology. [Note, the main purpose of COAG is to review and appraise issues in food and agriculture, and make recommendations on them to the FAO Council, which in turn reports to FAO's highest governing body, the FAO Conference]. The report of the Committee "stressed FAO's role in providing a forum for countries to monitor food and agriculture biotechnologies".

At its 116th session (14-19 June 1999), the FAO Council subsequently endorsed the COAG report, stating that it "appreciated the need for FAO to have a coherent programme on agricultural biotechnology to assist Member Nations in obtaining the full benefits of new developments while minimizing risks. FAO's role as a forum for the discussion of issues and for standard-setting, and as an 'honest broker' of quality science-based information, through mechanisms such as the International Plant Protection Commission (IPPC) and Codex Alimentarius, was underscored in general, and in relation to biotechnology in particular". Later, at the 30th session of the FAO Conference (12-23 November 1999), Members stated that one of the substantive areas to which they attached particular importance was the active contribution of FAO to current debates on biotechnology and genetically modified organisms. It was therefore in this spirit that FAO established the Forum. It is coordinated by the FAO Inter-Departmental Working Group on Biotechnology (IDWGB) that was established in 1999 following the recommendations of the 1999 COAG meeting.

Operation and Structure of the Forum

The Forum has an open structure that allows various parties — policy-makers, people from universities, NGOs, the public, etc. — to discuss and exchange views and experiences about specific issues concerning biotechnology and its applications in the animal, fishery, forestry and plant sectors in developing countries. The principal activity of the Forum is to run moderated e-mail conferences (each lasting roughly two months) about specific topics concerning biotechnology in food and agriculture for developing countries. To register for any conference, individuals must first be members of the Forum. The topics all have biotechnology as the core subject and may cover themes such as biosafety, public/private agricultural research, biodiversity, capacity-building, food safety, poverty alleviation, benefit sharing, intellectual property rights and food production. The emphasis is on developing countries. As the Forum covers the broad range of activities found within the area of food and agriculture, it covers topics both of specific relevance to those interested in the animal, crop, fish or forestry sectors or of general relevance to all sectors.

The Forum was officially launched on 9 March 2000. The launching was marked by sending an e-mail "letter of invitation" to a list of people and institutions that might have been interested in this initiative. An important source of e-mail addresses was a report on biotechnology networks in developing countries, prepared for FAO's Research and Technology Development Service (SDRR) in September 1999. The list also included Permanent Representatives to FAO and all FAO country representatives and was supplemented by additional addresses provided by members of the IDWGB. In the letter of invitation, people were requested to also forward the information to anyone that they considered might be interested.

The number of Forum members rose to over 700 within the first month, to over 1 000 after three months and to nearly 1300 by the time the sixth conference was finished more than one year later. Once they joined, very few people left the Forum. Forum members may not send messages to each other (although if they register for a conference they may send a message to all other participants in the conference) and may only receive messages from the Forum Administrator, who is responsible for all contact with the Forum members. They have so far (November 2001) received 31 messages i.e. roughly two a month from the Forum Administrator.

Forum members are not automatically registered for any e-mail conference, but instead have to do this themselves. Thus, they participated to varying degrees in the different conferences. Some did not register for any conference, but instead only received key documents from the Forum Administrator. Others instead, registered for several conferences and received all the e-mail messages posted.

When the Forum was launched, a website to complement and support the Forum was also launched (www.fao.org/biotech/forum.htm). The website was implemented in collaboration with FAO's information management group, WAICENT. Note, however, that the primary communication medium of the Forum is e-mail, so to be a Forum member and participate actively in any of its conferences, the only thing that is required is an e-mail account. The website merely gathers together in one place all the information about the Forum, as well as all the documents and individual messages related to the different conferences. The website has been recognized as a valuable resource. It was selected by the Internet Scout Project for inclusion in the Scout Report (26 May 2000), a weekly current awareness publication that highlights new internet resources of interest to researchers and educators; was chosen as a "Hot Pick" in the Netwatch section of the journal Science (www.sciencemag.org/cgi/content/summary/289/5479/503b) (28 July 2000); as well as the "site of the Day" by New Scientist (29 December 2000) (www.newscientist.com/weblinks/categories/agriculture2.jsp).

It was a conscious decision to operate the Forum with e-mail as the base communication medium (rather than, for example, running the conferences on the web) to try and facilitate participation from developing countries. Although both typically require access to a computer, modem, phone line and an account with an internet service provider, full internet access with browsing on the web tends to be more expensive and more difficult in practice than simply receiving and sending e-mail messages. The analyses carried out, showing that individuals from developing countries were actively involved in the Forum conferences but very seldom visited the Forum website, strongly support this decision.

Individuals wishing to join the Forum, have to register themselves. This is done by sending an e-mail message to an automatic FAO mail server. Using the server, people can automatically subscribe or unsubscribe themselves from the Forum, or they can receive messages previously posted by the Forum Administrator. Registration is also possible from the Forum website.

Operation of the Individual e-mail Conferences

The six conferences were operated in the same way.

a) Before a conference

Before a given e-mail conference began, a Background Document, two to five pages in length, was prepared. In this publication, the six Background Documents are included. As the conferences took place in a time span covering over one year, the documents were written at different stages from March 2000 to March 2001.

The aim of the Background Document is to give an easily-understandable description of the conference theme, enabling potential participants to have a basic grasp of some of the main aspects of the theme. For example, the Background Document to Conference 2, on the appropriateness of current biotechnologies for the forestry sector in developing countries, provided a brief summary of the kinds of biotechnologies currently available for the forestry sector; some key elements or current trends in the forestry sector in developing countries and finally, certain factors that should be considered in the discussions. Before a conference began, the Background Document was sent to Forum members. In the same e-mail message, they were invited to join the conference and given instructions about doing so. They were requested to carefully read the document if wishing to participate in the conference.

b) During a Conference

Involvement of Forum members in each conference is governed by the "Rules of the Forum" and "Guidelines for Participation in E-mail Conferences" that Forum members receive on joining the Forum. These specify, *inter alia*, that

- they should introduce themselves briefly in their first posting to a conference;
- they should exercise tolerance and respect toward other participants whose views may differ from their own, and remain courteous at all times;

- they should not submit messages longer than 600 words;
- people represent only themselves i.e. that "regardless of whether they identify the entity by whom they are employed, participants are assumed to be speaking in their personal capacity unless they explicitly state that their contribution represents the views of their organization. For this reason, participants should not quote the postings of other participants as representing the views of the organizations to which those other participants belong".

Each conference was moderated by the Forum Administrator. The Moderator's role is to screen all messages before they are posted to ensure that they follow the rules and guidelines of the Forum and that they are relevant to the theme of the conference. In addition, the Moderator plays an active role in the conference by ensuring that messages are understandable and, where appropriate, providing additional information of benefit to participants. Roughly 95 percent of messages received by the Moderator during FAO working hours were posted to the conference within an hour of receipt. Those received after working hours were usually posted first thing the following morning. Only a small minority of messages was refused for posting, coming mainly in Conference 1. Messages were refused primarily because they were not directly relevant to the theme of the conference. When required, IDWGB members provided technical support to the Moderator.

Midway through each conference, a brief Update Document was written and sent to Forum members, summarizing the kinds of messages posted, the subjects dealt with and pointing out some areas that should be addressed in the remaining time available. In some cases, as in Conference 1, more than one Update was written.

c) After a Conference

After a conference is finished, two Summary Documents are written. The first version is longer (five to eleven pages), more detailed and contains references to specific e-mail messages. The second version is shorter (one to two pages) and does not contain references. Both documents attempt to provide an easily-readable summary of the main arguments and concerns discussed during the conference, based on the messages posted by the participants. In this publication, the longer versions of the Summary Documents are provided. References are made to specific e-mail messages that can be viewed on the Forum website.

The Six Conferences

The first conference began less than two weeks after the Forum was launched. It was the first of a four-conference block on the theme of the appropriateness of currently available biotechnologies in the crop, forestry, animal and fishery sectors respectively for food and agriculture in developing countries. The themes of the fifth and sixth conferences were chosen based on the interest shown in them by participants during the early conferences. The fifth conference dealt with the implications of agricultural biotechnology for hunger and food security in developing countries, while the sixth examined the impact of intellectual property rights on food and agriculture in developing countries. The titles and start/end dates of the six conferences are as follows:

Conference 1 (20 March to 26 May 2000): How appropriate are currently available biotechnologies in the crop sector for food production and agriculture in developing countries?

Conference 2 (25 April to 30 June 2000): How appropriate are currently available biotechnologies for the forestry sector in developing countries?

Conference 3 (12 June to 25 August 2000): The appropriateness, significance and application of biotechnology options in the animal agriculture of developing countries.

Conference 4 (1 August to 8 October 2000): How appropriate are currently available biotechnologies for the fishery sector in developing countries?

Conference 5 (1 November to 17 December 2000): Can agricultural biotechnology help to reduce hunger and increase food security in developing countries?

Conference 6 (20 March to 14 May 2001): The impact of intellectual property rights (IPR) on food and agriculture in developing countries.

The conferences thus dealt with specific separate themes, although always remaining within the general area of biotechnology in food and agriculture in developing countries. The conferences attracted different audiences, discussed different topics (although they often overlapped) and each had different characteristics. Some general figures from the conferences are provided.

1.6 Limitations of the Conferences

(a) Language

The six conferences took place in English only. Thus the Background and Summary Documents for each conference, as well as all messages from the Moderator and by participants (with the exception of a couple of messages transmitted in both English and French in the livestock sector conference) were in English. This affects the kind of audience and participants in the different conferences, making it difficult for individuals lacking the English language to contribute to the conferences and to make their opinions/experiences known. For example, in the fisheries sector conference, there were no messages from some of the developing countries that have active programmes in fisheries biotechnology (Brazil, China and Cuba) and language might explain their absence. Nevertheless, messages posted in the Forum conferences came from nearly

50 different countries throughout the world, many of which do not have English as their main language.

(b) Electronic Communication

When people attend a "traditional" conference, the list of participants typically includes those who are invited and those who either pay for themselves or who are paid to attend by their employer. There is often a restriction on the maximum number of attendees. There is thus a certain process of selection where (language considerations apart) a number of people interested in a particular subject may not be present to provide their input. However, with any type of conference, the quality of the discussions and outputs depends on the participants. The aim of the Forum is to allow a wide range of parties to discuss and exchange views and experiences about specific issues concerning biotechnology. The medium for communication is e-mail and, as with a traditional conference, this also involves problems of selectivity.

Even though the Forum is free and open to everyone it requires, however, in the most typical case, electricity, a phone connection and a computer with a modem. There are thus large differences between and within countries regarding access to these new communication technologies. The UNDP Human Development Report 2001 (www.undp.org/hdr2001/) showed there is a large "digital divide" in the world today. Figures from the report indicated that, in the year 2000, nearly seven percent of the world population was using internet but that 79 percent of internet users lived in Organisation for Economic Co-operation and Development (OECD) countries. Furthermore, the percentage of the population with internet use ranged from 28 percent in high-income OECD countries to 0.4 percent in sub-Saharan Africa or South Asia. In addition, the report also provided some information on the "digital divide" *within* countries, pointing out that internet users are mainly:

- urban and located in certain regions (e.g. in China, only four million of 600 million people in 15 poorly connected provinces are internet users while two major cities, Shanghai and Beijing, have five million);
- better educated and wealthier (e.g. in Chile, 89 percent of users have third-level education);
- young (e.g. in China, 84 percent are under 35);
- male (e.g. 86 and 62 percent of users are male in Ethiopia and Latin America, respectively).

These are important limitations that should be kept in mind when reading the summaries from the conferences.

CROP SECTOR CONFERENCE

Background Document

Introduction

The biotechnology industry has developed in a very short time period to become a multi-billion dollar industry providing products for the areas of human health care, industrial processing, environmental bioremediation and food and agriculture. It is an industry that has developed, been financed and is firmly based in developed countries (especially North America). Whereas public funding for agricultural research has stagnated or declined, the biotechnology industry has continued to invest heavily in agricultural research due to the large advances made in the area and the strengthening of intellectual property rights for biological materials. The biotechnologies used and developed by the industry reflect market realities and are used primarily to provide products for developed countries. The biotechnologies used for food and agriculture are no exception in this regard. In this e-mail conference recently-developed biotechnologies that are currently available in the crop sector are discussed, in the context of how appropriate they are for food production and agriculture in developing countries.

Description of Currently Available Biotechnologies in the Crop Sector

It is probably fair to say that the most significant breakthroughs in recent years in the area of crop biotechnologies have stemmed from research into the genetic mechanisms behind economically important traits. The rapidly progressing discipline of genomics, providing information on the identity, location, impact and function of genes affecting such traits, is producing knowledge that has driven and will increasingly drive the application of biotechnologies in crops. Here, a summary of recently developed biotechnologies for the crop sector that could be used in practice for food production and agriculture in developing countries is provided.

Biotechnologies Based on Molecular Markers

All living things are made up of cells that are programmed by genetic material called DNA. This molecule is made up of a long chain of nitrogen-containing bases (A, C, G and T). Only a small fraction of the sequence in plants makes up genes, i.e. that code for proteins, while the remaining and major share of the DNA represents non-coding sequences whose role is not yet clearly understood. The genetic material is organized into sets of chromosomes (e.g. five pairs in the much-studied mustard species *Arabidopsis thaliana*) and the entire set is called the genome.

Molecular markers are identifiable DNA sequences, found at specific locations of the genome. They may differ between individuals of the same population. Different classes of markers exist, such as RFLPs, AFLPs, RAPDs or microsatellites.

Molecular markers can be used for:

- marker-assisted selection, which is the use of markers to increase the response to selection. A quantitative trait (i.e. one such as fruit yield that shows continuous variation and cannot be classified into a few discrete classes) is usually controlled by many genes, called quantitative trait loci (QTLs). By using molecular markers closely linked to, or even located within, one or more QTL, information at the DNA-level is used directly and selection response can be increased;
- marker-assisted introgression, where markers are used to increase the speed or efficiency of introgression (i.e. the introduction of new gene(s) from a population A to a population B by crossing A and B and then repeatedly backcrossing to B). Introgression may be of interest, for example, when wishing to introduce genes from wild relatives into modern plant varieties;
- studies of genetic diversity and of taxonomic/phylogenetic relationships between plant species or between populations (or varieties) within species;
- studies of biological processes, such as mating systems, pollen movement or seed dispersal, and of the genetic mechanisms behind physiological traits.

Genetically Modified Crops

Genetically modified organisms (GMOs) are those that have been modified by the application of recombinant DNA technology (where DNA from one organism is transferred to another organism). The term "transgenic crops" is also used for genetically modified crops, where a foreign gene (a transgene) is incorporated into the plant genome. It may help us to distinguish between three distinctive types of genetically modified crops:

- "Wide Transfer": where genes are transferred from organisms of other kingdoms (e.g. bacteria, animal) into plants;
- "Close Transfer": where genes are transferred from one species of plant to another;
- "Tweaking": where genes already present in the plant's genome are manipulated to change the level or pattern of expression.

Transgenic plants have been the subject of much controversy, although they now cover large areas in certain parts of the world. Estimates for 1999 indicate that 39.9 million hectares of land were planted with transgenic crops (ISAAA, 1999, www.isaaa.org/publications/briefs/Brief_12.htm). Of these, 7.1 (18 percent) were in developing countries, almost all in Argentina (6.7 million hectares) and China (0.3 million hectares), while the United States and Canada accounted for 32.7 million hectares (82 percent). Of the 39.9 million hectares, 28.1 million (i.e. 71 percent) were planted with crops modified for tolerance to a specific herbicide (which could be sprayed on the field, killing weeds while leaving the crop undamaged); 8.9 million hectares (22 percent) were modified to include a toxin-producing gene from a soil bacterium, *Bacillus thuringiensis*, which poisons insects feeding on the plant, while 2.9 million hectares (7 percent) were planted with crops having both herbicide tolerance and insect resistance.

Most of the transgenic crops planted so far have thus incorporated only a very limited number of genes. However, some transgenic crops of greater potential interest for developing countries have been developed in the research laboratories but have not yet been released commercially, such as transgenic rice of high iron content developed by transferring the ferritin gene from soybean to rice, or transgenic rice producing provitamin A.

Micropropagation

This is the in-vitro multiplication and/or regeneration of plant material under aseptic and controlled environmental conditions on specially prepared media that contain plant nutrition and growth regulators. The most commonly used materials are excised embryos, shoot-tips or pieces of stems, roots, leaves, etc.

It is the basis of a large commercial plant propagation industry involving hundreds of laboratories around the world. The technique can be used to multiply, in large numbers, clones of a particular variety. Apart from its rapid propagation advantages, micropropagation can also be used to generate disease-free planting material, especially if combined with the use of disease-detection diagnostic kits. Micropropagation techniques have been developed and are applied for a wide range of crops, including woody and fruit plants.

Food and Agriculture in Developing Countries

The emphasis of the e-mail conference is on developing countries. In this context, it should be kept in mind that a tremendous variety of production systems and environmental constraints are found between different developing countries and even within individual countries. Four broad agro-ecological zones (humid and peri-humid lowlands; hill and mountain areas; irrigated and naturally flooded areas; drylands and areas of uncertain rainfall) account for 90 percent of agricultural production in developing countries. Within each of the zones, a range of farming systems are found as well as a mixture of traditional and modern production systems.

The global population size has passed the six billion mark and is increasing by roughly 80 million annually. Almost all population growth is in developing countries. While the number of inhabitants in the developing and developed world, respectively, is estimated at 4.75 and 1.31 billion, respectively, for the year 2000, in 20 years time it is predicted to be 6.15 and 1.36 billion, respectively.

Farm sizes tend to be small, as reflected by a study of 57 developing countries showing that nearly 50 percent of farms were smaller than one hectare. The increase in food production needed to cover the increased population size cannot come from recruiting new land for agricultural purposes. Most land suitable for agriculture is already in use. When comparing the total amount of land of crop-producing potential with the amount of cultivated land, there are however noticeable differences between regions. For example, in South Asia, 191 of the potential 228 million hectares were already under cultivation in 1988-1990, whereas in Latin America and the Caribbean only 190 of the potential 1 059 million hectares were in use. However, parts of these could not be readily converted to crop production as they are already used for other purposes such as forestry, animal grazing or conservation. Degradation of land already in use, due to overgrazing, deforestation and poor farming practices, is also an increasing problem globally. The increases in food production needed to feed the world's growing population must therefore come from increasing the amount of food produced per hectare.

Note, however, that the issue of world hunger may not be simply solved by increasing the world food supply. In the world today enough food is produced to feed all its inhabitants but yet it is estimated that in 1995-1997 there were roughly 790 million undernourished people in developing countries, i.e. whose food intake was insufficient to meet basic energy requirements on a continuing basis (FAO, 1999, www.fao.org/NEWS/1999/991004-e.htm). Hunger and poverty are also influenced and determined by many different demographic, environmental, economic, social and political factors and these factors should also be considered when trying to reduce hunger in the world. Food needs to be available and accessible to the poor, wherever they may be.

Certain Factors that should be Considered in the Discussion

The key question in this e-mail conference is how appropriate each of the different biotechnologies, mentioned previously in this document, may be for the crop sector in developing countries and regions.

The question of appropriateness should consider the following elements:

- the factors determining or influencing the appropriateness of the different biotechnologies e.g. their environmental impact; their impact on human health; the status with respect to intellectual property rights; the status with respect to biosafety regulations and controls; the degree of access to the biotechnologies; the level of capacity-building or resources required to use them; their financial cost; their impact on food production and food security;
- the relative costs (financial, social, political or otherwise) of the biotechnologies versus the relative benefits (productivity, food security or otherwise);

- whether they are more (or less) appropriate than existing conventional methods in the crop sector for food production and agriculture, given the realities of life in developing countries;
- whether some of the biotechnologies are more (or less) appropriate than others;
- whether some biotechnologies are more (or less) suited to certain regions in the developing world than others.

Summary Document

The Background Document described three major kinds of recently developed biotechnologies that could potentially be used for the crop sector in developing countries: a) biotechnologies based on molecular markers; b) genetically modified (GM) crops; and c) micropropagation.

All three kinds of biotechnologies were discussed in the conference. However, the emphasis was overwhelmingly on GM crops. In some topics of discussion, messages representing strongly opposing points of view were posted, reflecting the polarization that exists regarding some elements of the debate on agricultural biotechnology.

In Section 2.2.1, some of the main factors that were discussed in the conference and considered to have direct importance for the appropriateness of the biotechnologies in developing countries are described. In Section 2.2.2, some other main arguments and concerns raised during the conference are described. In this document, references are included to specific messages. The participant's surname and the date posted (day/month of the year 2000) are provided. The messages can be viewed at www.fao.org/biotech/logs/c1logs.htm. In Section 2.2.3, the name and country of the participants that sent the referenced messages are provided.

Factors Considered of Direct Importance for the Appropriateness of Biotechnologies in Developing Countries

Their status with respect to intellectual property rights (IPR) and the potential power of multinational corporations (MNCs) as a consequence of IPR

The existence and impact of IPR over biotechnological products (e.g. plant varieties) and processes (e.g. techniques used in generating plant varieties) was probably the topic that attracted most discussion throughout the whole two-month long conference. The fact that a small number of powerful MNCs from developed countries had built up extensive patent portfolios meant that there was often a strong socio-political aspect to the discussion. Considerable differences of opinion were expressed about both the need for and consequences of IPR in the crop sector.

Some participants felt that IPR over biological materials were inherently wrong while others felt they were necessary. Berruyer (28/3 and 14/4) suggested it would be better if it was not possible to patent genes. Kumar (18/4) stated that the new seeds patented were developed from existing genetic material, often from developing countries, in a process involving very small (or no) genetic modification and so the patenting process converted something which was the "common heritage of mankind" into private property. She also argued that the process ignored the input over many generations from farmers in building up the base genetic material. Lettington (18/4) argued that enforcing IPR in developing countries created a net loss for humanity due to the lack of access to information.

On the other side, it was argued that farmers always have the choice as to whether or not to buy improved varieties from MNCs and that "those [companies] that invest in developing a product or technology should get paid for their creativity, capital risk-taking and simple hard work" (Laing, 17/4), a view that was also supported by Halos (4/4). Halos (17/5) suggested, in addition, that patenting genes did not mean that the major economic benefit went to the patent holder, but that many diverse groups, including farmers and consumers, also benefited from the GM varieties developed. Roberts (22/5) emphasized that business will only invest where it expects to make a profit and that in order for industry to invest in these technologies they should expect some financial return. Ashton (19/5) disagreed with this argument, maintaining that the nature of capitalism is that the developer bears the risk and nobody owes a return to the risk-taker.

The consequences of IPR were seen as being quite substantial. The point was made that the existence of strong IPR, and the fact that they are often owned by MNCs, would lead to (increased) dependence by developing country farmers on technologies owned by MNCs and developed countries. This was clearly expressed by Hongladarom (3/4) who indicated that "the fear [of biotechnology that has been aired in Thailand] does not so much concern the potential risks of the genetically modified crops as does the possibility that after a while farmers may have to rely exclusively on the technologies owned by these corporations". Berruyer (28/3) also made the same point saying "the problem with biotechnologies is not the tool, but who has the tool". Lettington (18/4) indicated that such dependency relationships were already being built up in East Africa. Salzman (24/3) feared that farmers in developing countries would be at the mercy of MNCs regarding pricing, seed supplies and the types of seeds provided. Reel (6/4) regretted the change by farmers from seed saving towards increased expense and dependence on outside seed resources. Schenkel (4/4), on the other hand, said that he did not see why farmers would become more dependent if the seeds were adapted to their needs.

Another consequence that was much discussed was that patents could be granted to companies from developed countries over genetic material from developing countries. Reel (6/4) provided information on specific examples, such as the yellow bean (Mexico) and basmati rice (India). Carneiro (13/4) pointed out that the recognition of IPR by developing countries opened up the possibility for developing countries to patent biotechnology products or processes either on their own or in joint projects. Munsanje (27/3), however, argued that developing countries lacked the financial resources required to "bioprospect" the large pool of biodiversity in their specific regions and to take economic and social advantage of their resources. Kumar (18/4) gave a concrete example of the problems raised by IPR, writing that each year in her country, Sri Lanka, many new tea and rice varieties are developed by national research institutes but they are never patented because the effective protection of a single variety in the major countries of the world would cost US$75 000-100 000. She noted, however, that there was nothing to prevent a private company patenting these varieties in the West and that government institutes would not be able to find the funds (maybe US$500 000 in the United States) needed to contest a patent. Ashton (19/5) said that measures to prevent "bio-piracy" were needed and that certain developments, such as the sale of some national seed banks in Africa to corporate interests, should be viewed with great concern.

The impact of IPR on plant breeding research in developing countries was also discussed. Carneiro (13/4) wrote that biotechnology research in developing countries was traditionally based on the transfer of technology but, following the adoption of IPR in developing countries, this approach was obsolete and therefore new products and processes specific for agriculture in developing countries had to be generated. Berruyer (14/4) argued that if patenting of genes was not allowed, then technology transfer would still be possible. Berruyer (14/4) also noted the difficulties of this new situation as developing countries now had to discover and develop the use of new genes, which is the most expensive part of the transgenic process and in addition, this had to be done in the context of competition from MNCs.

Some participants maintained that, in the light of this situation, MNCs had to take special consideration of developing countries. Fauquet/Taylor (26/5) proposed that MNCs should offer relevant technologies within their portfolios for use in developing country crops that do not represent a market to them in the near future. Olivares (12/5) proposed that, to encourage such measures, science policy in developed countries should support public science with the idea that the biotechnology products or processes obtained could be transferred free of charge to developing countries.

Others, instead, maintained that a new IPR system was needed. Munsanje (27/3) argued that IPR should be enhanced in developing countries in order to protect their products before they were exploited and patented. Lettington (18/4) argued that the whole current IPR system was developed in the North to serve a series of very particular purposes and that developing countries should develop their own parallel patenting system which would, for example, ensure that the holder of a patent on a traditional variety would compensate and recognize the developers of the variety.

Kumar (25/4) supported this view but felt that developed countries would strongly oppose the establishment of such a system.

Level of Resources or Capacity Building Required for their use in Developing Countries

It was argued that funds in developing countries are scarce and that often one of the first items in national budgets to be cut is 'research and development', making it very difficult for the countries themselves to develop biotechnology products that are suited to their own national needs (Nwalozie, 23/3; Halos, 23/3; Lettington, 24/3; Kuta, 30/3). Schenkel (22/5) emphasized that today the production of GM crops is still "very, very expensive".

Kiggundu (19/5) noted that third world governments typically do not have the finances to support conventional plant breeding activities and that, in this context, the availability of GM crops would be a breakthrough. However, Schenkel (22/5) argued that when there were insufficient resources to sustain conventional breeding, a country should not spend money on GM activities — a viewpoint strongly supported by Khan (22/5). Wingfield (13/4) noted that using biotechnologies in developing countries can be too expensive, especially when equipment has to be imported, and indicated that there was a definite niche available for people to develop procedures to apply biotechnology using locally available material.

Despite the lack of resources in many developing countries, Rebai (9/5) urged that, given the importance of agricultural biotechnology for food security, all developing countries "should keep trying to stay in the biotechnology train as drivers and not as spectators, as active makers and not passive consumers". Schenkel (22/5) also argued that the lack of resources should not mean that biotechnology would be exploited only by developed countries and that there was an obligation on developed countries to make biotechnology available to developing countries.

Their Impact on Human Health

There was much discussion regarding whether GM crops, in particular those producing toxins of the soil bacterium *Bacillus thuringiensis* (Bt), hereafter referred to as Bt-crops, could be harmful or allergenic (i.e. inducing allergies) when eaten by humans. Almost all contributions were from participants in developed countries. Large differences of opinion were expressed on this subject. Some participants maintained that they were at least as safe as non-GM food products while others argued that they were potentially highly allergenic. Some messages went into detail regarding testing procedures for allergenicity and, in some cases, links to websites providing further information were included.

Crystal proteins from Bt are toxins that kill insects feeding on the plant by binding to and creating pores in their midgut membranes. Both Reel (7/4) and Salzman (10/4) argued that there was no evidence that ingestion by humans of plants producing the toxin was safe. Roberts (10/4) stated that, based on the concept of "substantial equivalence", edible GM crops were tested in comparison with their non-modified counterparts and that, in general, no relevant differences in food quality were found and that neither the GM nor the non-GM plants were guaranteed to be "completely safe". Reel (3/4) pointed out that human testing, that might normally be carried out for a new food additive, was not required for GM foods and that testing them on animals (such as mice) was insufficient. Roberts (12/4) counter-argued that the digestive systems of humans were fundamentally different from those of insects and that results of testing with animals could be treated with confidence because of their close relationship to humans.

Berruyer (12/4) and Berruyer and Bucchini (in a joint message of 17/4) then provided more technical details regarding the working of the toxins, describing how most proteins, including Bt toxins, are denatured (i.e. the specific activity is destroyed) by the acidity of the human stomach. Bucchini, in the joint message (17/4), concluded that it is unlikely that the toxin endangers human health but urged caution. He argued (19/4) that there are no direct methods to assess the potential allergenicity of proteins from sources that are not known to produce food allergy. Berruyer, in the joint message (17/4), suggested that the risk of an allergic reaction that endangers human life is low and quite difficult to measure. De Kochko (13/4) argued that Bt had been used for years in organic farming and that "any product, absolutely any product and not only Bt toxin, can be allergenic for someone in particular. Bt toxin has not been shown to be more allergenic (and certainly less) than chocolate or peanut butter!!!"

Some specific concerns were expressed about Cry9C, one of the Bt toxins, which is heat- and digestion-resistant (Bucchini, 17/4; Berruyer/Bucchini, 17/4). The gene producing the toxin has been transferred to GM corn which has been under consideration for use as human food in the United States. Lin (18/4) argued that the fact that it had so far only been approved for animal feed and industrial uses (and not for human consumption) suggested that the regulatory system in the United States works.

Another specific product that was discussed was a transgenic soybean crop, developed as a potential animal feed, containing a gene transferred from the Brazil nut species that expresses a high-methionine protein. A study published in 1996 revealed that the protein was allergenic and Reel (7/4) suggested that this finding was a cause for concern regarding the cultivation of GM crops. Wingfield (10/4), on the other hand, argued that this showed that science works since the results were the consequence of efficient testing of the crop before release and that, from the results of the trials, the crops were found to be unacceptable and were not then used commercially.

Their Environmental Impact

As specified in the Background Document, of the estimated 39.9 million hectares planted with transgenic crops in 1999, 28.1 million (i.e. 71 percent) were modified for tolerance to a specific herbicide, 8.9 million (22 percent) were Bt-crops while 2.9 million (7 percent) were planted with crops having both herbicide tolerance and insect resistance. Most of the messages posted concerning the environmental impact of new biotechnologies dealt with Bt-crops.

a) Pest-resistant GM Crops

Some participants expressed the fear that large-scale planting of Bt-crops would accelerate the development of Bt resistance among pests. Geiger (24/3 and 4/4) was one of these, adding that in tropical areas, with several generations of pests per year, this would happen quickly. Reel (29/3) maintained that major companies in the field of agricultural biotechnology were aware that resistance was inevitable and were thus already developing successors to Bt-crops. Geiger (4/4) said that the loss of Bt as an insecticide would be a major loss for farmers and for society. Smith (27/3) counter-argued that the selection pressure on insects to develop resistance would not be any greater than with the use of chemical pesticides.

Another potential concern with Bt-crops (Lettington, 28/3; Srinivasan, 3/4) was raised by a study published in the scientific journal Nature on 2 December 1999 indicating that the Bt toxin exudes from the roots of Bt-corn and that it might therefore have negative consequences on soil ecosystems. Lin (4/4) emphasized that the authors could not indicate how the soil communities might be affected. Halos (17/5) suggested that these results from the laboratory were not supported by field experiments.

The positive impact on the environment of finding alternatives to the current large-scale usage of chemical insecticides was also discussed. Halos (24/3) wrote that corn farmers in the Philippines admit to using a lot of pesticides and that, until the possibility of Bt-corn arose, they saw no alternative. Srinivasan (3/4) reported from an FAO press release that global insecticide sales amounted to about US$12 billion in 1995; that more insecticides were used on cotton than on any other crop and that over two-thirds of the global cotton area treated with insecticides was in India, China and Pakistan. He argued that the introduction of Bt-cotton in these countries would be expected to reduce insecticide applications and their adverse environmental implications. Several other participants also said they expected that Bt-crops would lead to reduced insecticide use (e.g. Halos, 23/3; Açikgöz, 24/3; Smith, 27/3; Berruyer, 28/3; Bartsch, 31/3). However, there seemed to be disagreement about whether the Bt-crops grown so far had in fact resulted in such reductions. Lettington (3/4) cited a study on soybean crops where pesticide use was higher, while Smith (27/3) quoted from an American newspaper article indicating reductions in insecticide sales following use of Bt-corn.

Lettington (28/3) noted that both chemical insecticides and Bt-crops had some problems, such as development of resistance by the insects, and proposed that integrated pest management (IPM), although more time-consuming, might be preferable to GM crops. Halos (27/3) described the situation in the Philippines where corn farms tend to be no bigger than one hectare and, since farmers often have other jobs, she argued that they find IPM too time-consuming.

b) Herbicide-tolerant GM Crops

There was much less discussion about herbicide-tolerant crops than Bt-crops. Schestibratov (9/5) argued that GM crops resistant to non-selective herbicides (i.e. that kill almost all plants that are sprayed) meant that fewer and less-expensive herbicides could be used. Srinivasan (3/4) suggested that growing them resulted in an increased use of herbicides. The potential spread of herbicide resistance to other plant species was a cause for concern. Kumar (31/3) said that the development of a fast-growing herbicide-tolerant weed could have very serious implications in a small developing country. Berruyer (28/3) suggested that such GM crops should be forbidden in areas containing related wild species.

c) Impact on Biodiversity

It was suggested that biotechnology could have a positive impact on biodiversity in the environment, by increasing the amount of food produced per unit of land area and thus reducing the need to use forest or natural habitats for additional food production in the future (e.g. Paiva, 3/4; Wingfield, 6/4; Roberts, 12/4).

Regarding within crop species diversity, Laing (17/4) indicated that the increasing loss of diverse germplasm was a cause for concern. He said that the availability of improved varieties, often developed using new biotechnologies and producing higher yields, resulted in small-scale farmers neglecting their traditional varieties. Yibrah (25/5) also predicted that the use of GM crops, coming from a narrow genetic base, would lead to genetic erosion.

Their Status with Respect to Biosafety Regulations and Controls

It was suggested that the application and monitoring of biosafety regulations would be more difficult in developing than in developed countries. Thus, Kumar (31/3) wrote that "developing countries possess limited scientific infrastructure and expertise and do not have the wherewithal to monitor such experiments or the products of such experiments. Furthermore, they are ill equipped to deal with any environmental disasters emanating from these products." Sivaramakrishnan (14/4) argued that even in a country with a strong biosafety system in force, such as India, the monitoring process would not be very easy. Yibrah (25/5) maintained that the lack of finances would make it extremely difficult to assess or monitor GM crops. Ashton (19/5) said there had been insufficient consideration given to the ability of developing countries to cope with potential negative consequences and that those promoting the use of GM crops would not accept the risks which, in his country, would instead be borne by the farmers, retailers and consumers of South Africa. Lettington (28/3) emphasized the need for capacity building in developing countries in the area of biosafety.

Their Role as Tools to Increase Food Production, Food Security and to Reduce Hunger in Developing Countries

As indicated in the Background Document, the global population is increasing, the amount of land available is finite and more food per hectare is needed in the future, to avoid growing crops on land currently devoted to functions other than food production. Some participants felt therefore that biotechnology was an important element in this process (e.g. Lin, 30/3 and 31/3; Paiva, 3/4; Fauquet/Taylor, 26/5) and that it would help to maintain or increase food security in developing countries (Schenkel,16/5; Alexandratos, 16/5; Halos, 17/5).

Others argued that social and political factors were of greater importance (e.g. Lohberger, 31/3; Lettington, 3/4; Reel, 3/4), which could be seen by the fact that, even today, when sufficient food is produced globally, there is still hunger and poverty in many developing countries (Yibrah, 25/5). Some messages went a step further and suggested that, in some cases, pro-biotechnology parties argued that biotechnology could reduce world hunger for public relations purposes (Lettington, 3/4; Yibrah, 25/5).

Lin (31/3) and McGuire (31/3) emphasized that biotechnology alone could not solve the problem of world hunger but that it could contribute to solving it. McGuire also pointed out that "it is unrealistic (and unreasonable) to expect Southern agricultural scientists to become political activists as well, especially in charged settings". Reel (6/4) agreed that biotechnology researchers tended to be reluctant about getting involved in the politics and economics of their field, but argued that economic imperatives governed the benefits of their research.

Other Main Arguments or Recurrent Themes from the Conference

The conference was moderated and every effort was made to ensure that participants kept their contributions strictly to the subject of the conference, although in some cases this was difficult. Here, some of the other main or recurring themes from the conference are summarized.

The Relative Appropriateness of the Different Biotechnologies

This topic was addressed in several messages. Yibrah (25/5) insisted that developing countries should select the techniques that are most relevant to their own situations and priorities and that, in this context, MAS and micropropagation were more suitable than the development of GM crops. Srinivasan (12/4) maintained that the application of marker-based QTL studies had so far proved unsatisfactory, as it had resulted in few examples of new genetically improved varieties, especially for crops in developing countries. He agreed with comments in a 1996 scientific paper that this was due to the fact that QTL detection analyses and variety development were two different processes and that most QTL studies were directed towards elite genetic material.

Schenkel (12/4), using the example of a QTL analysis project in Indonesia which had limited success, argued that the marker-based approach might be limited because of the extensive field-testing required for QTL analysis and the large amounts of time and money required. This time aspect was also emphasized by Rebai (25/4) who indicated that it would take at least four years to develop improved varieties by MAS, whereas genetic modification could give improved varieties in one or two years. However, he also pointed out that MAS could do some of the same things as genetic modification and even more. Thus, for traits controlled by many genes, such as disease resistance, he suggested that MAS might be more useful than genetic modification.

Ashton (19/5) proposed that micropropagation was a more suitable technology for developing countries than genetic modification from a risk point of view and that many centres in Africa had developed capacity with micropropagation technology. He also argued that technologies involving molecular markers should not be emphasized as they were complex and not well understood. Wingfield (13/4) argued that micropropagation was a low-level technology that had tremendous benefits to offer for developing countries, citing the production of virus-free sweet potatoes in Zimbabwe as a good example. Loebenstein (29/3) also suggested that the combination of efficient virus assay procedures with rapid propagation technologies could have large advantages for sweet potato and the potato in developing countries.

Wingfield (13/4) mentioned that for cloning of eucalyptus trees in South Africa, cuttings were mostly used rather than micropropagation, due to costs. Halos (17/5) also agreed that micropropagation could be very useful in developing countries but added that, in her experience, labour and electricity were the major costs and thus the technology might only be profitable when the product involved is traditionally expensive, such as banana. Halos (17/5) considered that the use of DNA markers was still too expensive at this stage for breeders in developing countries.

Some participants (Guiltinan, 24/3; McGuire, 31/3; Wingfield, 3/4) highlighted the fact that genetic modification is not the only biotechnology available to the crop sector in developing countries. They argued that it represents just one of a suite of available technologies and that the often controversial debate on GM crops should not inhibit the use of other non-GM biotechnologies in developing countries.

Srinivasan (25/5) provided a reminder that there is also a regional or local dimension to the relative appropriateness of different biotechnologies: that more complex biotechnologies might be appropriate in high-producing regions while low-level technologies should be emphasized in areas of low productivity.

The Appropriateness of Different Biotechnologies for Different Parts of the Developing World

Lin (30/3) proposed that the appropriateness of different biotechnology products was a complex issue, often depending on factors specific to the country or region. Moscardi (28/3) said that it was useful to distinguish between two regions in Latin America and the Caribbean (LAC). The first, including countries located outside the tropical belt, is a more temperate region where modern technologies are available and well integrated with the agro-industry and where they have put together IPR and biosafety rules. In the second, including countries between the tropics, there is little application of biotechnology and little private or public sector investment in agricultural research.

Srinivasan (25/5) proposed that a division into high and low potential productivity regions would be useful. In the high producing areas, such as South Central China or Northwest India, biotechnologies should be developed both to maintain the existing high levels and to raise the yield ceilings. In the low producing areas, such as Southwest and Northeast China and parts of Africa, the emphasis should be on low-risk/low-cost biotechnologies such as micropropagation.

The appropriateness of new biotechnologies compared to conventional methods

Yibrah (25/5) said he was not convinced of the relative advantages of GM crops compared to conventionally improved or even local varieties. He proposed that for poor countries such as Uganda and Ethiopia it "may be better to rationally use the scarce resources available on more conventional, but appropriate technologies than advocating the use of GM crops". His views corresponded with those of Schenkel (4/4) who stated "I believe the cost effectivity of any technology should be the determining factor in developing countries. If there is an easy and cheap way to achieve a goal, first use this before applying the high tech expensive one!!!". Schenkel (4/4 and 22/5) argued that if there was a lack of basics — such as seed supply, extension services or breeding — then it was not appropriate to spend scarce resources on biotechnologies, since the best return from these resources would be got from conventional methods of agronomy and breeding.

Schenkel (12/4 and 22/5) also emphasized that molecular techniques should be applied within a sound conventional breeding programme, since strategies such as MAS cannot replace conventional breeding but merely supplement it and they can only be successful if an efficient breeding strategy is already in place. To use QTLs, he therefore proposed (12/4) that an efficient breeding programme be first established, that initial efforts should focus on single gene traits that are difficult to select under normal circumstances (e.g. sex determination in nutmeg, where farmers are unable to determine sex until flowering, which takes about 6-8 years (Srinivasan, 12/4) and that, having found markers for these traits, they should be used in national breeding programmes

Traits that are most Relevant for Improvement in the Crop Sector in Developing Countries

This topic was raised indirectly in many different messages, and was occasionally seen in a socio-political dimension. In the context of herbicide-tolerant GM crops, the potential benefits of selecting for labour-saving traits in developing countries were addressed. Lin (30/3) suggested that these crops would eliminate the use of labour for weeding and thus lower earning potential and poverty reduction in many instances, although in other sectors of developing countries where labour was scarce they would be advantageous (he contrasted this with insect resistance which he indicated was a desirable trait for both small and large-scale farmers in developing countries). Salzman (24/3) argued that labour in itself was not a bad thing and that farmers in developing countries would prefer to work on the land than to migrate to urban areas. Halos (27/3) argued that increasing the amount of labour in farms would not reduce poverty in her country, the Philippines. Smith (27/3) suggested that migration of the workforce from rural to urban areas is an inevitable feature of the economic maturation of a nation. Lettington (24/3) maintained that herbicide-tolerance would have little relevance in developing countries because most farmers would not be able to afford the herbicide.

Fauquet/Taylor (26/5) highlighted the fact that in developing the first generation of transgenic crops, scientists had considered traits, such as herbicide tolerance and insect resistance, which would be of interest within the economic realities of industrialized countries and that the products were never intended to address the needs of developing countries. Srinivasan (18/5) supported this view, saying that the current products were not directly relevant to the needs of small farmers in developing countries. The importance of developing biotechnology products that would address specific problems of developing countries (i.e. with improvements in the traits of major interest in these countries), rather than simply using those that are already available from developed countries, was underlined by several participants (e.g. Munsanje, 27/3; Lettington, 3/4; Wingfield, 3/4; Mwangi, 10/4). For example, Archak (22/5) noted that crops with improved salinity tolerance were keenly awaited in countries such as India.

There may be limits, however, to the traits that may be incorporated into the new biotechnology products. Kiggundu (19/5) argued that in his country, Uganda, there were serious agricultural problems due to factors such as land fragmentation, increasing population pressure and soil erosion and that GM crops with appropriate traits could help to alleviate these problems. Both Schenkel (22/5) and Yibrah (25/5), however, counter-argued that these kinds of problems would not be solved using GM crops but through changing detrimental agricultural practices and that investments in improving the extension services would be more worthwhile.

Polarization of the Biotechnology Debate and the Need for Balanced Information

When this FAO Biotechnology Forum was established, it was recognized that in some areas of agricultural biotechnology, the debate was quite polarized and it was hoped that, by providing a neutral forum for different parties to exchange views and experiences, this polarization might in some way be reduced because, in the words of Lettington (27/3), "as soon as the different interest groups refuse to talk and acknowledge each others concerns we are all in trouble". The large differences between the sides can be seen by comparing some of the messages posted in the conference. For example, both Reel (6/4) and Halos (17/5) consider the impact of GM crops on areas such as the environment, human health and society and come to totally opposing conclusions regarding their impacts and consequences, with numerous references provided from the scientific literature (both refereed and non-refereed) to back up their respective cases.

Some of the factors leading to this polarization were discussed. Salzman (22/5) suggested that polarization was inevitable as GM crops had been grown commercially without sufficient consultation and before there was a thorough investigation of the potential hazards and problems. Srinivasan (18/5) argued that recent developments in "terminator gene" technology, with MNCs claiming numerous patents in the area, had further polarized public opinion.

Archak (9/5) argued that polarization had implications at the farmer level, since government organizations were influenced by the political party in power while NGOs tended to oppose biotechnology. Thus, correct information about biotechnology rarely reached the farmers. The importance of the availability of good balanced information on a controversial topic such as GM crops was also emphasized in other messages. Knausenberger (15/5) said that fora such as this one, would help public understanding of the issues and that a publicly funded agency, such as FAO, should remain objective and not commit itself to any paradigm. Towards the end of the conference, Ashton (19/5) said that although many of the messages posted had reflected the polarity of the debate, it was "refreshing to see some meeting of minds. Dogmatism and polemic do little for the debate from either side but instead we should concentrate on the common ground."

It is obviously difficult to measure whether the conference had some impact on polarization. However, in the current "electronic age", e-mail conferences such as this can also reach audiences beyond the actual participants. We know, for example, that the conference was discussed in an article in the scientific journal Nature (1 June), it was used as the basis of an article in the Spanish national newspaper El País (19

July), referring especially to the message of Yibrah (25/5), and as research material for a series of Finnish television programmes on GMOs.

The use of Biotechnology in Developed Countries to Feed the Developing World

Alexandratos (15/5) argued that consideration of the welfare and food security in developing countries should not ignore the fact that they are net importers of food and that the amount imported, coming mainly from North America, Europe and Australia, had increased in recent years and was predicted to increase even further towards the year 2030. He suggested (16/5) therefore that the application of biotechnology in developed countries, to allow them to meet these expected export requirements, was thus important for food security in developing countries. Yibrah (25/5) rejected this line of argument and suggested that increased food production in countries such as Argentina and the United States and its cheap export could not solve the hunger and poverty problems in developing countries, since it did not address their cause — lack of fair trade and justice. Lettington (24/3), furthermore, suggested that the use of biotechnology in developed countries could have a negative impact on small farmers in developing countries, by increasing oversupply in developed countries and consequently depressing world prices.

GM Crops and Evolution

In transgenic crops, a foreign gene (or genes) is incorporated into the plant's genetic material. The gene may be from the same species, a related plant species or even a species from another kingdom (e.g. from winter flounder fish to the strawberry or from Bt to corn). The evolutionary implications of such across-species transfer of genetic material were discussed in a few messages.

Salzman (30/3 and 31/3) argued that crossing the species barriers is non-adaptive and contradicts the process of natural selection and that the creation of GM crops such as Bt-corn runs counter to the normal tendencies of nature and evolution (which tend to minimize the opportunities for crossing the species barrier) and so there is therefore the risk of a global ecological disaster. Knausenberger (15/5) expressed the same fears because "million of years of co-evolution are being circumvented". Both Schenkel (30/3) and Rebai (28/4 and 9/5) argued instead that crossing the species, genera and, sometimes, family barrier was something that happened naturally in nature (although rarely) or that could be achieved artificially. It was pointed out that some common food crops (such as bread wheat and canola) contained genetic material from more than one species and that some crops created by plant breeders and used already for many years were interspecies hybrids, such as triticale (a hybrid of *Triticum aestivum* and *Secale cereale*).

Public versus Private Sector

Lin (30/3) argued that whereas the "green revolution" was based on the results of scientific research carried out in public institutions, the new age of agricultural biotechnology was driven by tools developed and patented by private and not public, institutions and that a second "green revolution" would depend on a rethinking of the role of public research and on incentives to the private industry to make the tools available. McGuire (31/3) supported these views, emphasizing that the role of public research needed to be both rethought and revitalized. Carneiro (13/4) noted that investment in science and technology was much lower in developing than in developed countries and that the public research sector needed to find new ways of promoting scientific development in developing countries. He argued that there was a need to build up relationships between public and private sectors at both national and international levels, and between the scientific and production sectors. Fauquet/ Taylor (26/5) also emphasized the need for collaboration between the public and private sectors in developed countries with policy-makers, scientists, breeders, extension workers and farmers in developing countries. Berruyer (14/4), however, warned that cooperation between public research institutes in developing countries and powerful MNCs might be biased by foreign private interests and would not favour small farmers in developing countries.

Name and Country of Participants with Referenced Messages

- Açikgöz, Nazimi. Turkey
- Alexandratos, Nikos. Italy
- Archak, Sunil. India
- Ashton, Glenn. South Africa
- Bartsch, Detlef. Germany
- Berruyer, Romain. France
- Bucchini, Luca. United States
- Carneiro, Mauro. Brazil
- De Kochko, Alexandre. France
- Fauquet, C.M./Taylor, Nigel. United States
- Geiger, Chris. United States
- Guiltinan, Mark. United States
- Halos, Saturnina. The Philippines
- Hongladarom, Soraj. Thailand
- Khan, Iftikhar Ahmad. Pakistan
- Kiggundu, Andrew. Uganda
- Knausenberger, Walter. Kenya.

- Kumar, Vijaya. Sri Lanka
- Kuta, Danladi Dada. Nigeria
- Laing, Mark. South Africa
- Lettington, Robert. Kenya
- Lin, Edo. France
- Loebenstein, Gad. Israel
- Lohberger, Ben. Australia
- McGuire, Shawn. Netherlands
- Moscardi, Edgardo. Colombia.
- Munsanje, Elliot. Zambia
- Mwangi, Peter. Kenya
- Nwalozie, Marcel. Senegal
- Olivares, Jose. Spain
- Paiva, Edilson. Brazil
- Rebai, Ahmed. Tunisia
- Reel, Jeffrey. United States
- Roberts, Tim. United Kingdom
- Salzman, Lorna. United States
- Schenkel, Werner. Germany
- Schestibratov, Konstantin. Russia
- Sivaramakrishnan, Siva. India
- Smith, Jay. United States
- Srinivasan, Ancha. Japan
- Wingfield, Brenda. South Africa
- Yibrah, Haile Selassie. Ethiopia

FISHERY SECTOR CONFERENCE

Background Document

Introduction

Biotechnology in fisheries and aquaculture represents a range of technologies that present opportunities to increase growth rate in farmed species, improve nutrition of feeds for aquaculture, improve fish health, help restore and protect environments, extend the range of aquatic species and improve management and conservation of wild stocks. In this e-mail conference, the focus is on genetic biotechnologies, with a brief treatment of related reproductive and gene banking technologies, and the appropriateness of their application in the fishery sector in developing countries. It is important to note that developing countries produce more fishery products from aquaculture, inland capture fisheries and marine capture fisheries than developed countries. The coverage of the biotechnologies here is not comprehensive, but should be enough to stimulate discussion in the conference.

The vast majority of aquatic genetic resources are found in wild populations of fishes, invertebrates and aquatic plants. Fishstat, the FAO database on fishery statistics, lists 1 235 taxa of common aquatic species that are harvested by humans in major fisheries; thousands more species are taken by small-scale fishers. It also contains information on 440 species that are farmed, but just 20 of these taxa account for approximately 80 percent of world aquaculture production. Domestication of aquatic species has not proceeded to the same level as it has in the crop and livestock sectors. Genetic biotechnologies must be used both to assist in the further domestication of aquatic species and to help manage and conserve the genetic resources found in wild populations.

Genetic biotechnologies in the fishery sector

This Background Document provides a summary of recently developed biotechnologies that could be used or more widely used, in the fishery sector in developing countries. Genetic biotechnologies that can be used in fisheries and in aquaculture include those that help to manage genetic resources and those for genetic improvement.

For management of genetic resources, markers can be used in the identification of management units and of endangered species to assist fishery management and they can also help broodstock management in stocking programmes. These markers may be genes, proteins (i.e. the products of genes) sequences of DNA or the phenotypic expression of genes (different colours, shapes, etc.). In the 1960's, analysis of proteins revealed a wealth of genetic diversity in wild populations. Protein analysis is now relatively fast and inexpensive, but it requires tissue samples to be stored and transported frozen. DNA analysis is becoming the method of choice because of the small amount of tissue needed, the fact that the tissue can be stored dried or in alcohol and because DNA analysis reveals much more genetic variation than protein analysis.

Several kinds of DNA markers exist, such as RFLPs, AFLPs, RAPDs and microsatellites. These, as well as other kinds of markers, can be used to analyse gene frequencies and genetic variation in and between different groups of fish. Studies carried out using these technologies in fish populations have revealed high levels of genetic variation distributed throughout the fish genome.

Genetic improvement technologies cover a range of techniques requiring different levels of expertise and resources. Chromosome-set manipulation (i.e. polyploidy induction) is an established technique to increase the number of chromosome sets (ploidy number) in an organism. Temperature, chemical and pressure shocks applied to fish eggs can be used to produce triploid (three chromosome sets) individuals that have desirable culture traits. Sex-reversal and the production of single-sex groups of fish is also a simple technology that combines hormone treatment and chromosome-set manipulation.

Hybridization, i.e. the mating of genetically different groups from the same species (intra-specific hybridization) or from different species (inter-specific hybridization), is a simple technique that is now easy to accomplish due to our increased knowledge of reproductive biology. It can be used to combine good traits from two different species into one group of fish or to transfer a characteristic of one group to another. A problem is that breeding hybrids with hybrids results in a non-uniform and unpredictable group of fish that is generally not well suited for culture. Therefore, for hybrid production, the parent-lines must be maintained pure. The above genetic improvement techniques are considered short-term strategies, where the gains are seen in one or two generations.

Selective breeding is a longer-term strategy where gains are accumulated at each generation of selection. Molecular markers may now increase the efficiency of selective breeding by facilitating the identification of quantitative trait loci, i.e. genes that control complex characters such as growth rate and environmental tolerance and secondly, by making it possible to use molecular markers linked to QTLs to identify desirable individuals or families.

Genetic engineering and the production of transgenic organisms is an active area of research and development in aquaculture. This is a medium-long-term strategy in that development and testing of stable transgenic lines requires time. The large size and hardy nature of many fish eggs allows them to be manipulated rather easily and facilitates gene transfer by direct injection of a foreign gene or by electroporation, where an electric field assists gene transfer.

In the next three sections, currently available biotechnologies are briefly discussed in the context of fishery management, aquaculture and conservation respectively.

Biotechnologies in Fishery Management

The role that the application of genetic principles can play in the sustainable use and conservation of living aquatic resources is being increasingly appreciated by resource managers, policy-makers and the international community. Fishery management requires information on the fishery resources in order to be effective. Primary information needs include:

- an identification of the resource;
- the breeding or stock structure of the resource;
- an estimate of the size of the resource; and
- the identification of key habitat that the resource requires.

Genetic analysis of the resources can address these information needs. Gene and genotype frequencies of different markers can provide information on, *inter alia,* species identification, population stock structure, hybridization and gene flow. Often, data from other sources, e.g. studies of tagged fish or of external characters of fish, cannot provide such information or are extremely difficult to collect in certain areas such as large river systems, floodplains or marine areas.

The use of protein and DNA data in fishery management requires collection of baseline (or background) genetic information. Genetic data were used to determine how sub-groups of Pacific salmon differed from each other in the Pacific Northwest. This required the analysis of hundreds of stocks of salmon but, once completed, endangered stocks were identified, levels of migration were estimated and the contribution of different stocks to a mixed stock ocean fishery was estimated.

Protein and DNA information has been used to identify endangered species that are either inadvertently captured in wild fisheries or are purposefully taken illegally. DNA analysis of legally sold whale meat revealed that many samples came from protected species of whale and dolphin. Species of shark are often difficult to identify because it is only the fins or flesh that are for sale; DNA analysis can be used to identify the species that provided the tissue and has the added advantage that dried tissue or less-than-fresh samples from markets can be studied.

Biotechnologies in Aquaculture

Genetic biotechnologies in aquaculture focus primarily on increasing growth rate, but also include disease resistance and increased environmental tolerance. There are several biotechnologies that can be applied to farmed aquatic species.

Selective breeding, i.e. traditional animal breeding, started with the common carp several thousand years ago. However, it has only recently been applied to a handful of other species of food fish such as catfish, trout and tilapia. Therefore, many farmed aquatic species are very similar to their wild relatives. Selective breeding programmes have yielded significant and consistent gains of 5-20 percent per generation in species of, *inter alia,* Atlantic salmon, catfish and tilapia.

Hybridization is a simple genetic technology that has become easier with the development of artificial breeding techniques, such as the use of pituitary gland extract and other hormones to initiate gamete development and induce spawning (i.e.

the depositing of eggs), and an increased understanding of environmental cues that influence reproduction, such as day length, temperature or water current. Many of the natural reproductive isolating mechanisms that species develop in the wild can now be overcome by fish farmers.

These improvements in reproductive technologies have also assisted aquaculturists greatly in their efforts to domesticate aquatic species. In addition, by making it possible to remove the natural constraints and timing of breeding, farmers are able to mate many more species at the times that are most beneficial and thus help to ensure a steady and consistent supply of fish to the market.

Chromosome-set manipulation can be used to produce triploid organisms that generally do not channel energy into reproduction because of problems associated with development of reproductive organs. Initially it was thought that this energy saving would result in increased growth rate, but this seems not to be the case. The real advantage of triploids seems to be in their functional sterility. For example, triploid oysters do not produce gonads (i.e. reproductive glands) and are therefore marketable at times of the year when mature oysters have an off-taste because of production of gametes (i.e. sex cells — the ovum, or egg (female), and sperm (male)).

In aquaculture, one sex is often more desirable than the other. For example, female sturgeon produce caviar, male tilapia grow faster than females whereas it is the female trout and salmon that generally grow faster than the males. The production of single-sex groups of fish takes advantage of these differences between the sexes and can be accomplished by manipulation of the developing gametes and embryo. The manipulation can be in the form of denaturing (i.e. destroying) the DNA in gametes followed by chromosome-set manipulation or by hormonal sex-reversal and subsequent breeding. The phenotypic sex of many aquatic species can be changed by administering appropriate hormones. For example, genetically male tilapia can be turned into females through estrogen treatments. These genetic males when mated with normal males produce a group of all-male tilapia that grow faster and have less unwanted matings (that lead to overcrowding and stunting) than a group of mixed-sex tilapia. Some of the all-male offspring would have two male chromosomes and these could be used as broodstock for subsequent generations, thus avoiding the use of hormones in the broodstock. Hybridization can also be used to produce single-sex groups of fish, when the sex-determining mechanisms in the parental lines are different (for example, hybridization of Nile tilapia and the blue tilapia).

Genetic engineering is a vague term that has come to be nearly synonymous with gene transfer, i.e. the production of transgenic fish or GMOs. This technology is progressing rapidly and it is now possible to move genes between distantly related species. Gene transfer in fish has usually involved genes that produce growth hormone and has been shown to dramatically increase growth rate in carp, catfish, salmon, tilapia, mudloach and trout. A gene from the winter flounder that produces an anti-freeze protein was put into salmon in the hope of extending the farming range of the fish. The gene did not produce enough of the protein to extend the salmon's range into colder waters, but it did allow the salmon to continue growing during cold months when non-transgenic salmon would not grow. Transgenic technology is currently in the research and development stage. To our knowledge there are no transgenic aquatic plants or animals available to the consumer.

Cryopreservation

The development of cryopreservation or low-temperature technology allows the short- and long-term storage of gametes. Currently, these low-temperature techniques can only be used on male gametes; eggs and embryos can generally not be stored in this way. Freezing gametes can increase the flexibility of a fish breeder, especially when breeding species where the sexes mature or migrate at different times, when the breeding season is very short, when the breeders are far apart or when one sex is exceptionally rare.

Fish Health

Genetic biotechnologies are being used to improve fish health through conventional selection for disease resistance and through the use of molecular investigation of pathogens for characterization and diagnosis. DNA-based technologies are being used now to characterize different species and strains of pathogens. Genetic characterization of the pathogen may also reveal information about its origin, e.g. DNA analysis revealed two strains of crayfish plague fungus in Sweden: one from the local species and one originating in Turkey. Once the pathogen is characterized, DNA probes can be developed to screen for specific pathogens in tissue, whole animals and even in water and soil samples. These techniques are being used to detect viral diseases of marine shrimp throughout the world and for bacterial and fungal pathogens in fishes in many areas. Genetically engineered vaccines are also being developed to protect fish against pathogens. Genetic immunization of rainbow trout with a glycoprotein gene from the virus causing viral haemorrhagic septicaemia has recently been shown to induce high levels of protection against the virus. Work is also underway on immunizing carp, salmon and other fishes with genetically engineered vaccines for other diseases.

The new molecular techniques are extremely sensitive and can identify pathogens in fish long before there are any clinical signs of the disease. This has implications for quarantine and the trade of aquatic species, which is currently governed by the World Trade Organization and the Office International des Epizooties. Trade can be restricted based on the disease status of a product or a region; identification of minute quantities of a pathogen or of a new strain of an existing pathogen could change or influence existing trade patterns.

Farming System

Farming systems for aquatic species are diverse and include industrial-scale farms, family ponds and culture based fisheries (stocking), in both developed and developing countries. Often, there is a division of the production process where fingerlings (i.e. small fish, especially up to one year of age) or eggs are produced by the seed-supplier, but the grow-out to market size is done elsewhere. In the case of sea going salmon, there is often a seed supplier operating a hatchery near a river, a fingerling producer in a freshwater lake, and another group that grows the fish to market size in the sea. Marine shrimp hatcheries in Asia are usually small family-owned ventures, whereas in Latin America they are more industrial in scale. Appropriateness of genetic biotechnologies must take these different systems into consideration.

Biotechnologies in Conservation

Genetic biotechnologies can be used to reduce the impacts of farmed fish on wild populations, to identify and manage endangered species and to manage captive populations in aquaria or in species recovery programmes. In several areas, farmed fish must be made triploid, i.e. sterile, in order to reduce their impact on wild populations should they escape from the fish farm. Generally, the planned use of transgenic fish also includes the provision that they are sterile, to reduce the chance of mixing with other fishes. Genetic manipulation and polyploidization can be combined to regenerate endangered species. This can be done from frozen sperm by denaturing the DNA in an egg of a related species, fertilizing with frozen sperm from the endangered species and then duplicating the chromosome set of the fertilized egg.

Certain Factors that should be Considered in the Discussion

The key question in this e-mail conference is how appropriate the different biotechnologies may be for the fishery and aquaculture sectors in developing countries today.

The question of appropriateness should consider the following elements:

- how does the farming system influence the use of genetic biotechnologies in developing countries?
- what are the factors that determine or influence the appropriateness of the different biotechnologies in developing countries e.g. their environmental impact; their impact on human health; the status with respect to intellectual property rights; the status with respect to biosafety regulations and controls; the degree of access to the biotechnologies; the level of capacity-building or resources required to use them; their financial cost; their impact on food production and food security;

- the relative costs (financial, social, political or otherwise) of the biotechnologies versus the relative benefits (productivity, food security or otherwise);
- whether they are more (or less) appropriate than existing conventional methods in the fishery sector in developing countries;
- whether some of the biotechnologies are more (or less) appropriate than others;
- whether some biotechnologies are more (or less) suited to certain regions in the developing world than others.

Summary Document

In the Background Document prepared for this conference, a brief coverage of some main biotechnologies was provided. These included the use of protein or DNA markers, triploidization, sex-reversal, hybridization, selective breeding, freezing of male gametes, genetic modification of fish and finally, DNA-based technologies to diagnose and characterize fish pathogens and to develop vaccines. They were discussed in the context of three main areas: fishery management, aquaculture and conservation.

However, participants in the conference focused to a large degree on a single biotechnology, the use of genetic modification, in a single main area, aquaculture. Of the 26 messages posted during the conference, 19 dealt only with this theme. Apart from genetic modification, the technology of triploidization was also much discussed, but only in the context of its application to GM fish.

A range of factors (such as the impact on human health, the status with respect to intellectual property rights, the costs or capacity-building required) that might influence the appropriateness of the different biotechnologies were also mentioned in the Background Document. But again, one factor dominated the discussions: the potential ecological risk or environmental impact of GM fish.

Section 5.2.1 of this document attempts to summarize the main elements of the discussions. Specific references to messages posted, giving the participant's surname and the date posted (day/month of the year 2000), are included. The messages can be viewed at www.fao.org/biotech/logs/c4logs.htm. Section 5.2.2 gives the name and country of the people that sent referenced messages. No messages came from some of the developing countries (Brazil, China and Cuba) that have active programmes in fisheries biotechnology.

Topics Discussed in the Conference

The Nature of GM Fish

There was some basic disagreement about how different GM fish were from non-GM fish. Muir (1/9) maintained that GM fish were very different as they could retain all the benefits of the wild species, while the transferred gene (the transgene) could potentially confer major advantages on the individual fish, such as being able to spawn at different times or invade new habitats. Conversely, the transgene could also make individuals less fit than wild types by affecting traits such as juvenile survival (Muir, 30/8). Moav (4/9 and 28/9) maintained that GM fish lines were similar to the domesticated parental lines which created them and that their genetic superiority for traits such as growth rate or disease resistance would be similar to that achievable through many years of conventional selective breeding.

Production of GM fish in Developing Countries

Currently, there is no commercial growth of GM fish, either in the developed or developing world. Norris (23/8), however, predicted that within the next five years or so, production of GM fish for human consumption would be a reality. She argued that there were two reasons why it might happen in a developing country such as Chile. The first is that Chile is an important producer of farmed fish, thus representing a major potential market for the technology. Secondly, consumer opposition to GMOs in general is far tamer than in developed countries, a point also made by Mair (15/9). Halos (12/9) emphasized that in densely populated developing countries with rising population numbers the priority is providing people with food as "poor people do not care to save for tomorrow since they fear tomorrow may not come for them, anyway". Mair (15/9) concluded that concerns about human health and environmental aspects of GM fish would inevitably be weighted lower when food security was a major issue, which could result in GM fish in aquaculture being adopted first in developing rather than developed countries.

Potential Environmental Impact of GM Fish

As mentioned earlier, this was the major topic taken up by the participants during the conference. Discussions touched on four main areas:

- the potential impact when GM fish are introduced into ecosystems where the wild species already exists;
- the potential impact when GM fish are introduced into ecosystems where the wild species does not exist;
- whether triploidization (and thus sterilization) of the GM fish would reduce the ecological risk;
- biosafety in developing countries.

a) Growing GM Fish where Wild Relatives Exist

Muir (30/8) pointed out that, unlike the situation of domesticated animals, domesticated GM fish might escape into an ecosystem where the wild non-GM members of the same species are found (e.g. a hypothetical case might be production of transgenic

Atlantic salmon in the Atlantic Ocean). He argued that this was a major concern because a) the wild relatives are likely to be an integral part of the ecosystem and disruption of the species could affect the entire ecosystem; and b) the escaped individuals can establish themselves by interbreeding with the wild relatives.

For this situation, Muir (30/8 and 1/9) summarized results from a paper he co-authored in 1999 which, using a theoretical model, considered the potential consequences of a small number of GM fish escaping and mating with their wild relatives. His results showed that if the transgene increased mating success but reduced the viability of transgenic offspring, then the local fish population could be driven to extinction. Halos (31/8) pointed out that the introduction of a new fish strain or of a superior conventionally bred strain might have the same consequences on a wild fish population and that this phenomenon was thus not unique to GM fish. She reported that this had already happened with the native catfish strain in the Philippines.

Regarding fish escapes, Halos (12/9) and Mair (15/9) both described the problems, especially due to large environmental extremes, of enforcing risk management strategies in developing countries. Mair (15/9) concluded, based on his practical experiences, "I would never like to guarantee that any domesticated fish cannot escape from an aquaculture facility".

Halos (31/8) argued that if GM fish mated with wild relatives, this might increase genetic diversity in the wild population. Muir (1/9), however, refuted this, concluding that GM fish (or exotic non-GM fish species) might in the short-term add diversity but, in the long-term, they decrease it because they eliminate competitors.

b) Growing GM Fish where Wild Relatives do not exist

Moav (4/9) pointed out that in his country, Israel, carp had been imported from Europe and that the transgenic carp (with increased growth rates) that they had developed in Israel would not present such potential problems as there was no native carp population. Muir (5/9) suggested that the issue of the production of GM fish in regions where the wild species does not exist was of great importance and that there was a range of other potential examples, such as the production of transgenic tilapia in Cuba or transgenic Atlantic salmon in the Pacific. He had two major concerns about such potential initiatives:

- he argued that the introduction of exotic non-native species usually resulted in ecosystem disruption and so should be treated with extreme caution — an argument that would be valid for both GM and non-GM introduced species. A much-quoted example of the potential hazards was the introduction of the (non-GM) grass carp, *Ctenopharyngodon idella*, species from Asia to the United States to control aquatic weeds (Kapuscinski, 22/9). The species caused much ecological damage and its escape into new ecosystems in the country has exacerbated the problem (Muir, 7/9). In this context, Ashton (25/9) also noted that the movement of African fish species between regions had caused serious environmental disruption to aquatic systems in Africa. Halos (12/9) underlined that the increasing need for food in developing countries has been a driving force behind the practice of introducing exotic fast-growing aquatic species. She argued that if GM fish gave higher yields per unit area and at a lower cost, then they might be a better alternative;

- Muir's (5/9) second concern was that such initiatives might give a false sense of security, as there was nevertheless a danger of the GM fish being later transported and introduced to other regions of the world where wild relatives are found — something which might be driven by economical imperatives. Mair (15/9) supported this point saying that "you can be fairly sure that if a fish is considered superior for aquaculture, its movement will be impossible to control completely". Muir (5/9) wondered who would be liable for any environmental damage in these situations.

c) Triploidization

With such concerns expressed about the potential ecological risk of GM fish mating with non-GM wild relatives, the potential application of triploidization to GM fish to ensure their sterility was raised (Ibarra, 6/9). Benfey (6/9) pointed out that reliable technologies for making GM fish triploid exist for salmonids and that this was a simple way to ensure they would not breed if they escaped into the wild. He also suggested that companies producing transgenic fish might want to only sell sterile fish, in order to protect their investment.

In theory, each individual GM fish could be tested to ensure it was triploid before being released, a procedure already established in some situations with the grass carp in the United States (Benfey, 7/9; Kapuscinski, 22/9). Chevassus (11/9), however, pointed out that it is possible to test for triploidy but not for sterility and that in some species, although not in salmonids, a few or large number of triploid individuals may in fact be fertile. Muir (6/9) also argued that it is actually hard to quantify how successful a sterilization technique such as triploidization may be if the true probability of failure is very low (e.g. one in a million), because, to reliably quantify it, an extremely large number of fish, more than normally tested, may be required.

Muir (11/9) also pointed out that even though triploid males might be sterile they may still mate with fertile females of the wild species, thus interfering with reproduction and breeding of the wild population. To avoid this potential problem, he thus proposed that GM fish, in addition to being made triploid, should also be sex-reversed so that only females be grown. Mork (11/9) reported that a Working Group on the Application of Genetics in Fisheries and Mariculture, belonging to the Mariculture Committee of the International Council for the Exploration of the Sea, had considered the issue of triploidization at various times throughout the 1990s. An impetus for this work was the finding that some previously triploid Pacific oysters (*Crassostrea gigas*) introduced to the east coast of the United States reverted back to the diploid state. Their conclusion in a 1995 report was that "no current mass triploidization/sterilization technique is guaranteed 100 percent effective".

Mair (15/9) pointed out that there was an additional reservation about the application of triploidization in aquaculture of GM fish in developing countries, i.e. that "the application of triploidy in commercial stocks (mainly salmonids and grass carp) has been limited to species that are habitually bred using artificial fertilization and incubation. For most of the important species in developing country aquaculture (namely tilapias and carp) artificial fertilization is rarely used and therefore application of triploidy on a commercial scale would be very unlikely to be viable".

d) Biosafety

Such discussions on the potential impact of GM fish escaping into the wild and the use of technologies such as triploidization to minimize potential risks, brought up the main issue of biosafety which, broadly defined in relation to GMOs, involves assessing and monitoring the effects of possible gene flow, competitiveness and the effects on other organisms, as well as possible deleterious effects of the products on health of animals and humans. Ibarra (6/9) noted that in developing countries there was a substantial lack of human resources in the fishery sector trained in genetics and that this could lead to the situation where "potentially high-risk biotechnologies will become implemented without a careful evaluation". Norris (23/8) also expressed the fear that GM fish might be introduced in developing countries "without even considering risk assessment for such introductions". Ashton (25/9) insisted that, prior to release of any fish, GM or not, in developing countries, there was a need for adequate biosafety protocols, legal instruments, liability procedures and a clear thread of responsibility for any damage that might be caused to the countries by their release. Del Valle Pignataro (27/9) lamented the fact that, in relation to introducing non-native species (GM or not) to developing countries, it was not possible in most cases to establish strict regulatory/monitoring systems, due to factors such as low economic priority or the lack of qualified human resources.

Gjoen (5/9) argued that it was difficult to foresee all the risks involved with GM fish and that the precautionary principle should be given priority, a view that was shared by Ashton (25/9). The need for carrying out risk assessment in a scientifically sound manner was emphasized in a few messages (Moav, 4/9; Muir, 5/9; Gjoen, 5/9; Moav, 28/9).

Use of Genetic Modification versus Other Alternatives

Genetic modification dominated discussions in the conference. Nevertheless, some participants did consider other biotechnologies and other aspects of aquaculture in developing countries. Doering's (25/9) perspective was that, with few exceptions, the fish species cultured today are wild and that enormous gains for traits such as productivity, growth rate or survival can be achieved by selective breeding, assisted by molecular methods. He argued that, apart from concerns about the potential environmental impact, "transgenic aquatic animals are not sensible or cost-effective in the genetic background of a wild animal and the enormous productivity gains to be made by intensive selective breeding". Norris (23/8) also emphasized that many developing countries were "in need of practical help and advice in developing good aquaculture breeding and husbandry practices which would benefit their programmes greatly".

Ibarra (6/9) suggested that most of the currently available genetic biotechnologies are very appropriate for developing countries and that the main reason for their under-use was the "lack of human resources within the fishery and aquaculture sector trained in the adequate use of those genetic biotechnologies".

Doering (25/9) emphasized that many of the current problems in aquaculture in developing countries have low-technology solutions and that "the species appropriate for culture in developing countries generally do not have the production economics to justify many high cost inputs such as vaccines and artificial larval feeds". He argued that policy-makers and scientists can become over-enthusiastic for molecular techniques, ignoring the large capacity-building needs that these technologies require. His conclusion was that "investments in developing countries on farmer education, reducing culture stress and improving water quality as well as domestication will yield higher returns than investments in high technologies".

Ashton (25/9) argued for the prioritization of local solutions in developing countries and that management systems which secure the protection, husbandry and sustainability of native species should first be put in place before any fish, GM or not, are introduced. Del Valle Pignataro (27/9) supported this viewpoint. She suggested that prioritization should be given to domestication, culture and (eventual) genetic improvement of native fish species that are already exploited and that have good consumer acceptance in developing countries. She gave a summary of their ongoing marine fish efforts in this direction in Mexico, which will eventually involve the use of selective breeding with medium-level biotechnologies

Name and Country of Participants with Referenced Messages

- Ashton, Glenn. South Africa
- Benfey, Tillmann. Canada
- Chevassus, Bernard. France

- Del Valle Pignataro, Gabriela. Mexico.
- Doering, Don. United States
- Gjoen, Hans Magnus. Norway
- Halos, Saturnina. The Philippines
- Ibarra, Ana. Mexico
- Kapuscinski, Anne. United States
- Mair, Graham. Thailand
- Moav, Boaz. Israel
- Mork, Jarle. Norway
- Muir, Bill. United States
- Norris, Ashie. Ireland

Glossary

Abiotic: Non-living matter or factors found in the environment (e.g. soil, altitude, slope)

Acequia (crop science): An irrigation ditch or canal.

Acre-foot (crop science): The volume of water that would cover one acre to a depth of one foot.

ADG (animal science): Average daily gain. The amount of body weight gain in a day.

Agribusiness: Producers and manufacturers of agricultural goods and services, such as fertilizer and farm equipment makers, food and fiber processors, wholesalers, transporters, and retail food and fiber outlets.

Agricultural ecosystem: An ecosystem where agriculture is the main contributor; a closed or confined habitat within an agricultural environment

Agronomy: Production of crops in a farming enterprise

AI (animal science): Artificial insemination. Impregnating an animal through artificial means, not through natural breeding.

Annual: Any plant which completes its entire life cycles and dies within one year or less.

Aquaculture: The production of products using water as the main source of growing medium

Aquifer A stratum of earth or permeable rock that stores significant quantities of water.

Associated Milk Producers, Inc. (AMPI) (dairy science) A farmer-owned milk cooperative that purchases milk from New Mexico dairy producers. Southern Region headquarters in Arlington, Texas.

AUM (animal science): Animal Unit Month. The amount of forage required by a mature cow (or other grazing animal) and her calf for one month.

Barrow (animal science): A male pig castrated before reaching sexual maturity.

BFP(dairy science): Basic formula price, determined by the U.S. Department of Agriculture, that serves as a basis for pricing milk in the United States. The BFP is calculated monthly based on previous BFP and the price of cheese.

Biological control: Controlling plants, diseases, and animal pests using natural enemies; or inhibiting the reproduction of pests by methods that result in the laying of infertile eggs, etc.

Biological diversity(wildlife science): Richness and abundance of species, and variety of natural communities. Both the number of species and the number of individuals within each species are important in considering the extent of biological diversity in an area. Also referred to as biodiversity.

Biotic: Living organisms on the planet Earth (e.g. vegetation, humans, bacteria)

Boar (animal science): A sexually mature male hog.

Breeding stock (animal science): Sexually mature male and female livestock that are retained to produce offspring.

Browse (forestry, wildlife science): Leaves, buds, and woody stems used as food by deer and other animals.

Brush: Commonly refers to undesirable shrubs and small trees.

BST (dairy science): Bovine somatotropin, commonly referred to as growth hormone. Produced naturally by the cow, stimulates metabolic functions related to growth and milk production.

Bull (animal science): An uncastrated male bovine.

Calf crop (animal science): The number or percentage of calves produced in a herd within a given year relative to the number of cows and heifers exposed to breeding.

Canopy (forestry): The uppermost layer in a forest, formed collectively by tree crowns.

Chisel (crop science): A farm implement used to break through and shatter compacted or otherwise impermeable layers of soil.

Class 1 milk (dairy science): Milk that is used for fluid use.

Clearcut harvest (forestry): A harvest and regeneration technique removing all the trees (regardless of size) on an area. Clearcutting is commonly used with shade-intolerant species such as Douglas fir or lodgepole pine, which require full sunlight to reproduce and grow well. Clearcutting produces an even-aged stand of trees.

Combine (crop science): A self-propelled or tractor-drawn machine which cuts, threshes, and cleans the standing crop which moving across the field. It is adapted to harvesting all the small grains, soybeans, grain sorghums, peanuts, beans, etc. In some areas, the crop is cut and placed in windrows by a swather (windrower), and a combine with a pickup attachment gathers the grain and threshes it at a later date.

Commercial farming: Specialised farming enterprise that is capital-intensive and aimed at profit maximisation crop rotation production of crops alternatively on the same piece of soil in consecutive seasons

Contour farming (crop science): Field operations such as plowing, planting, cultivating, and harvesting on the contour, or at right angles to the natural slope to reduce soil erosion, protect soil fertility, and use water more efficiently.

Cord (forestry): A stack of wood that has a gross volume of 128 cubic feet. A standard cord measures 4 feet x 4 feet x 8 feet.

Cover (wildlife science): Vegetation or other natural shelter serving to conceal wildlife from predators.

Crossbreeding (animal science): The mating of animals of different breeds. For example, breeding a Hereford cow with an Angus bull.

Cull (animal science): To select inferior animals from the herd for potential sale.

Cultivar: The name officially given for all cultivated varieties of plants.

Cultivator (crop science): A machine used to till the upper portion of the soil, primarily used to destroy weeds or form a moisture retaining mulch.

Deciduous tree (forestry): A tree that loses its leaves or needles during the fall and winter.

DHIA(dairy science): Dairy Herd Improvement Association. A nationwide system of performance testing and records processing for dairy producers.

Disk (crop science): An tractor-drawn implement composed of circular plates arranged at an angle with the soil. Used to prepare the soil for seeding.

Ditch rider (crop science): Manages water for an irrigation district. This person is responsible for delivering water through ditches to farmers for irrigation.

Dormant / dormancy: A biological process in which a plant ceases most growth activities and simply maintains existing tissue.

Dowsing (water witching): Using a forked stick or wire to locate underground water.

Drill (crop science): A farm implement for planting seeds which forms a small furrow, deposits the seed in dribbles, covers the seed, and packs soil over it. It can also deposit fertilizer, lime, or other amendments into the soil, alone or with the seed.

Drip irrigation (crop science): A method used to place irrigation water near plants' roots through pipes or tubes. This reduces water evaporation and runoff, but may not be cost effective for all crops.

Dry cow (dairy science): A cow that is not lactating.

Dry period (dairy science): A rest period between lactations when a cow is not lactating, normally 60 days in duration.

Dryland farming (crop science): Farming on non-irrigated land. Success is based on rainfall, moisture-conserving tillage, and drought-resistant crops.

El NiÃ±o: A recurring weather phenomenon where ocean temperature rises, resulting in a change of the upper air circulation and changing rainfall patterns all over the world

Entomology: The study of insects and their environments.

Environmental Impact Statement (EIS): A document evaluating the probable consequences of a proposed project that might significantly alter the environment. Required by the National Environmental Policy Act (1969) for any such project a U.S. government agency plants to undertake, regulate, or fund. An EIS is released in draft form (DEIS) to other agencies and the public for comment and review.

Ephemeral stream: A stream that flows only sporadically, such as after storms.

Estrus (heat)(animal science): The recurrent, restricted period of sexual receptivity in livestock. Non-pregnant cows and heifers usually come in heat 18 to 21 days following their previous estrus.

Evergreen tree(forestry): A tree that retains some or most of its leaves, or needles, throughout the year.

Ewe(animal science): A female sheep of any age.

Exotic species: A non-native plant or animal species introduced by humans, either deliberately or accidentally.

Fallow (crop science): Cropland left idle during the growing season.

Fed (or fat) cattle(animal science): Cattle that have been finished, usually in the feedlot, and are ready for slaughter.

Feeder cattle (animal science): Cattle past the calf stage that have weight increased making them salable as feedlot replacements.

Feedlot (animal science): A confinement facility where cattle are fed to produce beef for the commercial trade. May be under a roof or outdoors.

Field corn(crop science): Any variety of corn that is grown extensively in large fields primarily for livestock feed, as contrasted with the horticultural varieties, such as sweet corn or popcorn. Most field corn is of the dent variety.

Filly (animal science): A female horse less than three years old.

Firebreak (forestry): An existing barrier, or one constructed before a fire occurs, from which all or most flammable materials have been removed.

Fleece (animal science): The wool coat of a sheep.

Foal (animal science): A young horse of either sex, less than one year old.

Forage (animal science): Herbaceous plants or plant parts fed to domestic animals.

Forbs (animal science, wildlife science): Green broadleaf weeds.

Forestry: The science, art, and practice of managing and protecting tree and forest resources for human benefit.

Fungicidesee "pesticide"futures contract (agricultural economics): An agreement between two people one who sells and agrees to deliver, and one who buys and agrees to receive a certain kind, quality, and quantity of products to be delivered during a specified delivery month at a specified price.

Gelding (animal science): A castrated male horse.

Genetic modification: The manipulation of DNA structures of plants and animals to enhance productiongreenhouse structure (covered in glass or plastic) used to control the environment in order to increase productionherbivore an animal that utilises plants as its main source of food

Gilt (animal science): Sexually mature female hog, prior to having her first litter.

Green manure (crop science, gardening): A crop planted with the intention of turning it under for use as organic matter.

Groundwater: Water within the earth that supplies wells and springs.

Hardwood (forestry): A term describing broadleaf trees, usually deciduous, such as oaks, maples, cottonwood, ashes, and elms.

Heifer (animal science): A young female bovine.

Herbicide: see "pesticide"

Home range (wildlife science): The area within which the activities of an animal are confined. The boundaries may be marked, such as by scent marking, and also may be defended, depending on the species.

Homogenization (dairy science): The process of physically reducing the particle size of fat in milk, thus enabling even distribution of fat throughout the milk.

Horticulture: Production of vegetables and flowers

Hydroponics: Production of crops where soil is not used as the growing medium

Hypothesis: A proposed explanation or conclusion based on evidence

Indigenous: Plants or animals endemic to South Africa

Integrated pest management (IPM): Using the best features of chemical, biological, and cultural controls in an overall pest control program.

Lactation (animal science, dairy science): The time between calving that a cow produces milk, usually 10 months.

Lamb (animal science): A young sheep, less than one year old.

Land-grant universities: State colleges and universities started from federal government grants of land to each state to encourage further practical education in agriculture, home economics, and the mechanical arts.

Legume (crop science): A family of plants, including many valuable food and forage species, such as peas, beans, soybeans, peanuts, clovers, and alfalfas. They can convert nitrogen from the air to build up nitrogen in the soil.

Livestock: Any animal that is farmed to generate an income or food

Mastitis (dairy science): An infection and inflammation of the udder in cows.

Mid America Dairymen (Mid Am) (dairy science): A farmer-owned milk cooperative that purchases milk from New Mexico dairy producers. Headquarters in Missouri.

Milking parlor (dairy science): A facility that functions to extract milk from a cow by means of vacuum.

Milo (crop science): A small drought-resistant grain sorghum with compact bearded heads of large yellow or whitish seeds.

Mutton (animal science): Meat from sheep that are over one year old.

Nematicide: see "pesticide"

Net farm income (agricultural economics): The money and non-money income form operators realize from farming as a return for labor, investment, and management after production expenses have been paid. Net farm income is measure in two ways net farm income before inventory adjustment and net farm income after inventory adjustment. Net farm income doesn't include changes in the value of inventories such as crops and livestock at the end of the year.

Oil seed crops(crop science): Primarily soybeans, peanuts, cottonseed, and flaxseed used for the production of oils for cooking, protein meals, and nonfood uses. Less common oil crops include sunflower, safflower, castor beans, and sesame.

Open (animal science): A term commonly used to refer to a non-pregnant female.

Optimum production: Production that maximises returns on utilised inputs without exploiting resources

Organic farming: Production of crops and livestock without the use of non-organic chemicals

Parturition: The process of giving birth in farm animals

Pasteurization (dairy science): The process of heating milk to kill pathogens that may be harmful to humans.

Pasture (animal science): A fenced area of forage, usually improved, on which animals are grazed.

Perennial (crop science): A plant that lives for more than two years.

Permaculture: Permanent agriculture production of combining crops and animals

Permanent pasture (animal science): Pasture of perennial or self-seeding annual plants maintained through several years of grazing.

Pesticide: Used to destroy pests. Fungicides (destroys fungus), herbicides (destroys plants), insecticides (destroys insects), and nematicides (destroys nematodes) are all pesticides.

Photosynthesis(crop science, forestry): The process plants use to change air and water into food, using the sun's energy.

Polled (animal science): Naturally hornless cattle; having no horns.

Posilac (dairy science): Trade name for rBST marketed by Protiva, a unit of Monsanto.

Power-take-off (PTO) (crop science): A powered shaft, usually extending from the rear of the tractor and driven by the tractor motor, to supply rotative power to an attached or trailing implement such as a combine, hay baler, mower, etc.

Progeny (animal science): The offspring of animals.

Protein supplement (animal science): feed containing protein, vitamins, and minerals which is fed to livestock to provide a complete diet.

Ram (or buck) (animal science): A male sheep of any age.

Rangeland (animal science): A large open land area on which livestock wander and graze. The native vegetation is mainly grasses, forbs, and shrubs.

RBST (dairy science): Recombinant BST, produced by genetically modified bacteria. Increases milk yield of dairy cows when administered properly.

Riparian: Pertaining to the area along the banks of a river, stream, or lake.

Rotation (or crop rotation) (crop science): The growing of different crops, in recurring succession, on the same land.

Row crops (crop science): The rows or planting beds are far enough apart to permit the operation of machinery between them for cultural operations.

Ruminant: An animal that has a compound stomach (i.e. consisting of four portions of stomach)

Section: A surveyed parcel of land containing 640 acres one square mile.

Seedstock(animal science, plant science): Pedigreed or well-bred livestock which is maintained for breeding purposes. A specially selected strain of plants or seeds which are to be used as parents for future generations.

Selection harvest (forestry): Harvesting individual trees or small groups or trees at periodic intervals (usually 8 to 15 years) based primarily on their vigor and age. Selection harvesting perpetuates and uneven-aged stand.

Silage (animal science): A crop that has been preserved in a moist, succulent condition by partial fermentation in a tight container (silo) above or below ground. The chief crops stored in this way are corn (the whole plant), sorghum, and various legumes and grasses. The main use of silage is in cattle feed.

Sire (animal science): The male parent. To father or become the sire of.

Sorghum (crop science): A cereal grass used mainly for feedgrain or silage.

Sow (animal science): A sexually mature female hog, after having her first litter.

Soybeans (crop science): A legume crop, native to the Orient, used mainly in the United States for high protein feed and oil.

Stag (animal science): A male bovine castrated after puberty.

Stallion (animal science): An unaltered (uncastrated) male horse.

Stand: A recognizable area of plants that is relatively homogeneous and can be managed as a single unit.

Steer (animal science): A male bovine castrated early in life, usually as a calf.

Stockers (or stocker cattle) (animal science): Heifers and/or steers that are being grown on pasture or other forage for later sale as feedlot replacements.

Subsistence farming: Small-scale farming for household consumption sustainable agriculture production that avoids exploitation of natural resources while ensuring the maintenance of productivity

Total mixed ration (animal science, dairy science): A diet where all the feed ingredients are blended together to ensure every bit is nutritionally balanced.

Viticulture: Wine production weathering process of the breaking down of rocks to form soil over a period of time.

Bibliography

Acock, B. and Allen, L.H. Jr. 1985. Crop responses to elevated carbon dioxide concentration. In: *Direct Effects of Increasing Carbon Dioxide on Vegetation.* DOE/ER-0238. B.R. Strain and J.D. Cure (eds.). US Dept. of Energy, Carbon Dioxide Res. Div., Washington DC. pp. 53-97.

Adams, J. L. 1981. "Pies on the Prairies," Frontier, May, pp. 15-18.

Adams, R.M., Rosenzweig, C., Peart, R.M., Ritchie, J.T., McCarl, B.A., Glyer, J.D., Curry, R.B., Jones, J.W., Boote, K.J. and Allen, L.H. Jr. 1990. Global climate change and US agriculture. *Nature* 345: 219-224.

Aldwell, C. R., D. J. Burdon, and M. Sherwood. 1983. "Impact of Agriculture on Groundwater in Ireland," Environmental Geology, 5:39-48.

Alfoldi, L. 1983. "Topic 2: Movement and Interaction of Nitrates and Pesticides in the Vegetation Cover-Soil Groundwater-Rock System," Environmental Geology, 5:19-25. Alford, H. G., and M. P. Ferguson (eds.). 1982. Pesticides in Soil and Groundwater: Proceedings of a Conference, Division of Agricultural Sciences, University of California, Berkeley.

Allen GJ, Kuchitsu K, Chu SP, Murata Y, Schroeder JI (1999) *Arabidopsis abi1-1* and *abi2-1* phosphatase mutations reduce abscisic acid-induced cytoplasmic calcium rises in guard cells. Plant Cell 11: 1785-1798

Allen, L.H. Jr. 1990. Plant responses to rising carbon dioxide and potential interactions with air pollutants. *J. Environ. Qual.* 19: 15-34.

Allen, L.H. Jr. 1991. Effects of increasing carbon dioxide levels and climate change on plant growth, evapotranspiration, and water resources. In: *Proceedings of a Colloquium on Managing Water Resources in the West Under Conditions of Climatic Uncertainty.* 14-16 Nov. 1990, Scottsdale, AZ. National Research Council, National Academy Press, Washington DC. pp. 101-147.

Allen, L.H. Jr. 1994. Carbon dioxide increase: Direct impacts on crops and indirect effects mediated through anticipated climatic changes. In: *Physiology and Determination of Crop Yield.* K.J. Boote, J.M. Bennett, T.R. Sinclair and G.M. Paulsen (eds.). American Society of Agronomy, Crop Science Society of America, and Soil Science Society of America, Madison, Wisconsin. pp. 425-459.

Allen, L.H. Jr. and Amthor, J.S. 1995. Plant physiological responses to elevated CO2, temperature, air pollution, and UV-B radiation. In: *Biotic Feedbacks in the Global Climatic System: Will the Warming Increase the Warming?* G.M. Woodwell and F.T. Mackenzie (eds.). Oxford University Press, New York. pp. 51-84.

Allen, L.H. Jr., Baker, J.T., Albrecht, S.L., Boote, K.J., Pan, D. and Vu, J.C.V. 1995. Carbon dioxide and temperature effects on rice. In: *Climate Change and Rice.* S. Peng, K.T. Ingram, H.-U. Neue, and L.H. Ziska (eds.). Springer-Verlag, Berlin, Heidelberg. pp. 256-277.

Allen, L.H. Jr., Bisbal, E.C., Boote, K.J. and Jones, P.H. 1991. Soybean dry matter allocation under subambient and superambient levels of carbon dioxide. *Agron. J.* 83: 875-883.

Allen, L.H. Jr., Boote, K.J., Jones, J.W, Jones, P.H., Valle, R.R., Acock, B., Rogers, H.H. and Dahlman, R.C. 1987. Response of vegetation to rising carbon dioxide: Photosynthesis, biomass, and seed yield of soybean. *Global Biogeochemical Cycles* 1: 1-14.

Allen, L.H. Jr., Jones, P. and Jones, J.W. 1985. Rising atmospheric CO2 and evapotranspiration. In: *Advances in Evapotranspiration. ASAE Pub. 14-85.* American Society of Agricultural Engineers, St. Joseph, Michigan. pp. 13-27.

Allen, L.H. Jr., Valle, R.R., Mishoe, J.W, Jones, J.W. and Jones, P.H. 1990. Soybean leaf gas exchange responses to CO2 enrichment. *Soil Crop Sci. Soc. Fla. Proc.* 49: 192-198.

Allen, L.H. Jr., Vu, J.C.V., Valle, R.R., Boote, K.J. and Jones, P.H. 1988. Nonstructural carbohydrates and nitrogen of soybean grown under carbon dioxide enrichment. *Crop Sci.* 28: 84-94.

Allen, R.G. and Gichuki, F.N. 1989. Effects of projected CO2-induced climatic changes on irrigation water requirements in the Great Plains States (Texas, Oklahoma, Kansas, and Nebraska). In: *The Potential Effects of Global Climate Change on the United States.* Appendix C, Agriculture, Vol. 1. EPA-230-05-89-053. J.B. Smith and D.A. Tirpak (eds.). US Environmental Protection Agency, Washington DC. pp. 6-1 to 6-42.

Altieri, M.A. 1987. Agroecology: the scientific basis of alternative agriculture. Boulder: Westview Press.

Altieri, M.A. 1991. Traditional Farming in Latin America. The Ecologist 21:93-96.

Altieri, M.A. 1992. Agroecological foundations of alternative agriculture in California. Agriculture, Ecosystems and Environment 39: 23-53.

Altieri, M.A. 1994. Biodiversity and pest management in agroecosystems. Hayworth Press, New York.

Altieri, M.A. 1995. Agroecology: the science of sustainable agriculture. Westview Press, Boulder

Altieri, M.A. 1996. Enfoque agroecologico para el desarrollo de sistemas de produccion sostenibles en los Andes. Centro de Investigacion, Educaccion y Desarrollo, Lima.

Altieri, M.A. and C.I. Nicholls. 1999. Biodiversity, ecosystem function and insect pest management in agricultural systems. In W.W. Collins and C.O. Qualset (eds.) Biodiversity in Agroecosystems. CRC Press, Boca Raton.

Altieri, M.A. and D.K. Letourneau. 1984. Vegetation diversity and insect pest outbreaks. CRC Critical Reviews in Plant Sciences 2: 131-169.

Altieri, M.A. and D.L. Letourneau. 1982. Vegetation management and biological control in agroecosystems. Crop Protection 1: 405-430.

Altieri, M.A. and L. C. Merrick. 1987. In situ conservation of crop genetic resources through maintenance of traditional farming systems. Economic Botany 4:86-96.

Altieri, M.A. and L. C. Merrick. 1987. Peasant Agriculture and the conservation of crop and wild plant resources. J. Cons. Biol. 1: 49-58.

Altieri, M.A. and P.M. Rosset 1995. Agroecology and the conversion of large-scale conventional systems to sustainable management. International Journal of Environmental Studies 50: 165-185.

Altieri, M.A. and S.B. Hecht 1991. Agroecology and small farm development. CRC Press, Boca Raton.

Altieri, M.A., D.K. Letourneaour and J.R. Davis. 1983. Developing sustainable agroecosystems. BioScience 33: 45-49.

American Farmland Trust. 1986. "Confusion Hampers Conservation Reserve," American Farmland, 6(5):1.

Amthor, J.S. 1991. Respiration in a future, higher-CO2 world. *Plant Cell Environ.* 14: 13-20.

Amthor, J.S. 1995. Plant respiration responses to elevated partial CO2 pressures. In: *Advances in Carbon Dioxide Effects Research.* L.H. Allen, Jr., M.B. Kirkham, C. Whitman and D.M. Olzyk (eds.). Special Pub., American Society of Agronomy, Madison, Wisconsin.

Andow, D.A. 1991. Vegetational diversity and arthropod population response. Annual Review of Entomology 36: 561-586.

Anklin, M., and thirty-nine other Greenland Ice-core Project (GRIP) members. 1993. Climate instability during the last interglacial period recorded in the GRIP ice core. *Nature* 364: 203-207.

Arnaud J.-F. F. Viard M. Delescluse J. Cuguen 2003 Evidence for gene flow via seed dispersal from crop to wild relatives in *Beta vulgaris* (Chenopodiaceae): consequences for the release of genetically modified crops with weedy lineages. *Proceedings of the Royal Society of London, B, Biological Sciences* 70: 1565-1571

Arp, W.J. 1991. Effects of source-sink relations on photosynthetic acclimation to elevated CO2. *Plant Cell Environ.* 14: 869-875.

Audirac, Y. 1997. Rural sustainable development in America. John Wiley and Sons, N.Y.

Baker, D.N. and Enoch, H.Z. 1983. Plant growth and development. In: *CO2 and Plants.* E.R. Lemon (ed.). Westview Press, Boulder, Colorado. pp. 107-130.

Baker, D.N., Hesketh, J.D. and Duncan, W.G. 1972. Simulation of growth and yield in cotton: I. Gross photosynthesis, respiration, and growth. *Crop Sci.* 12: 431-435.

Baker, J.T. and Allen, L.H. Jr. 1993a. Contrasting crop species responses to CO2 and temperature: Rice, soybean, and citrus. *Vegetation* 104/105: 239-260. Also: pp. 239-260. In: *CO2 and Biosphere.* (Advances in Vegetation Science 14). J. Rozema, H. Lambers, S.C. van de Geijn and M.L. Cambridge (eds.). Kluwer Academic Publishers, Dordrecht.

Baker, J.T. and Allen, L.H. Jr. 1993b. Effects of CO2 and temperature on rice: A summary of five growing seasons. *J. Agric. Meteorol. (Japan)* 48: 575-582.

Baker, J.T. and Allen, L.H. Jr. 1994. Assessment of the impact of rising carbon dioxide and other potential climate changes on vegetation. *Environ. Pollut.* 83: 223-235.

Baker, J.T., Allen, L.H. Jr. and Boote, K.J. 1992a. Response of rice to CO2 and temperature. *Agric. For. Meteorol.* 60: 153-166.

Baker, J.T., Allen, L.H. Jr. and Boote, K.J. 1992b. Temperature effects on rice at elevated CO2 concentration. *J. Exp. Bot.* 43: 959-964.

Baker, J.T., Allen, L.H. Jr. and Boote, K.J. 1995. Potential climate change effects on rice: Carbon dioxide and temperature. In: *Climate Change and Agriculture: Analysis of Potential International Impacts. C.* Rosenzweig, L.A. Harper, S.E. Hollinger, J.W. Jones and L.H. Allen, Jr. (eds.). ASA Special Pub. No. 59, American Society of Agronomy, Madison, Wisconsin. pp. 31-47.

Baker, J.T., Allen, L.H. Jr., Boote, K.J. and Pickering, N.B. 1996. Assessment of rice responses to global climate change: CO2. and temperature. In: *Terrestrial Ecosystem Response to Elevated CO2*. G.W. Koch and H.A. Mooney (eds.). Physiological Ecology Series, Academic Press, San Diego. pp. 265-282.

Baker, J.T., Allen, L.H. Jr., Boote, K.J., Jones, J.W. and Jones, P. 1990a. Developmental responses of rice to photoperiod and carbon dioxide concentration. *Agric. For. Meteorol.* 50: 201-210.

Baker, J.T., Allen, L.H. Jr., Boote, K.J., Jones, P. and Jones, J.W. 1989. Response of soybean to air temperature and carbon dioxide concentration. *Crop Sci.* 29: 98-105.

Baker, J.T., Allen, L.H. Jr., Boote, K.J., Jones, P. and Jones, J.W. 1990b. Rice photosynthesis and evapotranspiration in subambient, ambient, and superambient carbon dioxide concentrations. *Agron. J.* 82: 834-840.

Baker, J.T., Allen, L.H. Jr., Boote, K.J., Rowland-Bamford, A.J., Jones, J.W, Jones, P.H., Bowes, G. and Albrecht, S.L. 1988. Response of vegetation to carbon dioxide, Ser. No. 043. Response of rice to subambient and superambient carbon dioxide concentrations, 1986-1987. Report of the US Dept. of Agriculture, Agricultural Research Service (in cooperation with the University of Florida) for the US Dept. of Energy, Carbon Dioxide Research Division, Office of Energy Research, Washington DC.

Baker, J.T., Laugel, R, Boote, K.J. and Allen, L.H. Jr. 1992c. Effects of daytime carbon dioxide concentration on dark respiration of rice. *Plant Cell Environ.* 15: 231-239.

Barles, B., and J. Kotas. 1987. "Pesticides and the Nation's Ground Water," EPA Journal, 13(4):42-43.

Bartsch D, Lehnen M, Clegg J, Pohl-Orf M, Schuphan I, Ellstrand NC (1999) Impact of gene flow from cultivated beet on genetic diversity of wild sea beet populations. Mol Ecol 8: 1733-1741.

Batie, S. S. 1983. Soil Erosion: Crisis in America's Croplands? The Conservation Service Washington, DC.

Batten M. 1983. "The Rush Is On to Study Jungles," International Wildlife, 13(3):16-19.

Baulcombe DC (1999) Fast forward genetics based on virus-induced gene silencing. Curr Opin Plant Biol 2: 109-113.

Bazzaz, F.A., Jasienski, M., Thomas, S.C. and Wayne, P. 1995. Microevolutionary responses in experimental populations of plants in CO2-enriched environments. Parallel results from two model systems. *Proc. of the National Academy of Sciences* S.92: 8161-8165.

Beachy RN (1997) Mechanisms and applications of pathogen-derived resistance in transgenic plants. Curr Opin Biotech 8: 215-220 .

Beets, W.C. 1982. Multiple Cropping and Tropical Farming Systems. Westview Press, Boulder.

Beets, W.C. 1990. Raising and sustaining productivity of smallholders farming systems in the Tropics. AgBe Publishing, Holland.

Beever, D E, Kemp C F; 2000; Safety issues associ- ated with DNA in animal feed derived from geneti- cally modified crops: a review of scientific and reg- ulatory procedures; Nutrition Abstracts and Review Series B: Livestock feeds and feeding. 70(3), pp 175-182

Bertrand, A. 1981. "Biological Diversity: Basic for Agricultural Success,"

Blauert, J. and S. Zadek 1998. Mediating sustainability. Kumarian Press, Connecticut.

Boote, K.J., Pickering, N.B., Baker, J.T. and Allen, L.H. Jr. 1994. Modelling leaf and canopy photosynthesis of rice in response to carbon dioxide. *International Rice Research Notes* 19: 47-48.

Broecker, W.S. 1987. Unpleasant surprises in the greenhouse? *Nature* 328: 123-126.

Brokenshaw, D.W., D.M. Warren and O. Werner. 1980. Indigenous Knowledge Systems and Development. University Press of America, Lanham.

Brookfield, H. and C. Padoch 1994. Appreciating agrobiodiversity: a look at the dynamism and diversity of indigenous farming practices. Environment 36: 7-20.

Brush, S.B. 1982. The natural and human environment of the central Andes. Mountain Research and Development 2:14-38.

Bullock J. M. 1999 Using population matrix models to target GMO risk assessment. *Aspects of Applied Biology* 53: 205-212

Bunce, J.A. 1990. Short- and long-term inhibition of respiratory carbon dioxide efflux by elevated carbon dioxide. *Ann. Bot.* 65: 637-642.

Burke, D; 2002; The UK regulatory process in prac- tice; in GM Crops: understanding the issues; Agricultural Biotechnology Council; pp 61-68

Busch, L., W.B. Lacy, J. Burkhardt and L. Lacy 1990. Plants, Power and Profit. Basil Blackwell, Oxford.

Buttel, F.H. and M.E. Gertler 1982. Agricultural structure, agricultural policy and environmental quality. Agriculture and Environment 7: 101-119.

Caballero, J.N. and C. Mapes 1985. Gathering and subsistence patterns among the P'urepecha Indians of Mexico. J. Ethnobiol. 5: 31-47

Camilleri C, Lafleuriel J, Macadre C, Varoquaux F, Parmentier Y, Picard G, Caboche M, Bouchez D (1998) A YAC contig map of *Arabidopsis thaliana* chromosome 3. Plant J 14: 633-642 .

Campbell, W.J., Allen, L.H. Jr. and Bowes, G. 1988. Effects of CO2 concentration on rubisco activity, amount, and photosynthesis in soybean leaves. *Plant Physiol.* 88: 1310-1316.

Campbell, W.J., Allen, L.H. Jr. and Bowes, G. 1990. Response of soybean canopy photosynthesis to CO2, light, and temperature. *J. Exp. Bot.* 41: 427-433.

Carrol, C. R., J.H. Vandermeer and P.M. Rosset. 1990. Agroecology. McGraw Hill Publishing Company, New York.

Chambers, R. 1983. Rural Development: putting the last first. Longman Group Limited, Essex.

Chang, J.H. 1977. Tropical agriculture: crop diversity and crop yields. Econ. Geogr. 53:241-254.

Clarke, W.C. and R.R. Thaman 1993. Agroforestry in the Pacific: systems for sustainability. United Nations university Press, Tokyo.

Clawson, D.L. 1985. Harvest security and intraspecific diversity in traditional tropical agriculture. Econ. Bot. 39:56-67.

Colwell R. E. E. A. Norse D. Pimentel F. E. Sharples D. Simberloff 1985 Genetic engineering in agriculture. *Science* 229: 111-112[Free Full Text]

Conner A. J. T. R. Glare J.-P. Nap 2003 The release of genetically modified crops into the environment. II. Overview of ecological risk assessment. *The Plant Journal* 33: 19-46.

Conway GR (1999) The Doubly Green Revolution: Food for All in the 21st Century. Cornell University Press, Ithaca, NY

Conway, G.R. and Pretty, J.N. 1991. Unwelcome harvest: agriculture and pollution. Earthscan Publisher, London.

Crawley M. J. S. L. Brown R. S. Hails D. D. Kohn M. Rees 2001 Transgenic crops in natural habitats. *Nature* 409: 682-683.

Crosson, P. 1993. Impacts of climate change on the agriculture and economy of the Missouri, Iowa, Nebraska, and Kansas (MINK) region. In: *Agricultural Dimensions of Global Climate Change.* H.M. Kaiser and T.E. Drennen (eds.). St. Lucie Press, Delray Beach, Florida. pp. 117-135.

Cure, J.D. and Acock, B. 1986. Crop responses to carbon dioxide doubling: A literature survey. *Agric. For. Meteorol.* 38: 127-145.

Curry, R.B., Peart, R.M., Jones, J.W., Boote, K.J. and Allen, L.H. Jr. 1990a. Simulation as a tool for analyzing crop response to climate change. *Trans. ASAE* 33: 981-990.

Curry, R.B., Peart, R.M., Jones, J.W., Boote, K.J. and Allen, L.H. Jr. 1990b. Response of crop yield to predicted changes in climate and atmospheric CO2 using simulation. *Trans. ASAE* 33: 1383-1390.

Curry. R.B., Jones, J.W., Boote, K.J., Peart, R.M., Allen, L.H. Jr. and Pickering, N.B. 1995. Response of soybean to predicted climate change in the USA. 1995. In: *Climate Change and Agriculture: Analysis of Potential International Impacts.* C. Rosenzweig, L.A. Harper, S.E. Hollinger, J.W. Jones, and L.H. Allen, Jr. (eds.). ASA Special Pub. No. 59, American Society of Agronomy, Madison, Wisconsin. pp. 163-182.

Daniell H. 2002 Molecular strategies for gene containment in transgenic crops. *Nature Biotechnology* 20: 581-586.

Darwin C. 1876 The effects of cross and self fertilization in the vegetable kingdom. John Murray, London, UK

Davis GL, McMullen MD, Baysdorfer C, Musket T, Grant D, Staebell M, Xu G, Polacco M, Koster L, MeliaHancock S, Houchins K, Chao S, Coe EH (1999) A maize map standard with sequenced core markers, grass genome reference points and 932 expressed sequence tagged sites (ESTs) in a 1736-locus map. Genetics 152: 1137-1172.

De la Fuente-Martinez JM, Herrera-Estrella L (1999) Advances in the understanding of aluminum toxicity and the development of aluminum-tolerant transgenic plants. Adv Agron 66: 103-120.

Della-Cioppa G, Grill LK (1996) Production of novel compounds in higher plants by transfection with RNA viral vectors. Ann NY Acad Sci 792: 57-61.

Delucia, E.H., Sasek, T.W. and Strain, B.R. 1985. Photosynthesis inhibition after long-term exposure to elevated levels of atmospheric carbon dioxide. *Photosyn. Res.* 7: 175-184.

Den Nijs H. D. Bartsch J. Sweet [eds.] 2004 Introgression from genetically modified plants into wild relatives and its consequences. CABI, Wallingford, UK

Denevan, W.M. 1995. Prehistoric agricultural methods as models for sustainability. Advanced Plant Pathology 11:21-43.

Denevan, W.M., J.M. Treace, J.B. Alcorn, C. Padoch, J. Denslow and S.T. Paitan. 1984. Indigenous agroforestry in the Peruvian Amazon: Bora Indian management of swidden

Desprez T, Amselem J, Caboche M, Hofte H (1998) Differential gene expression in *Arabidopsis* monitored using cDNA arrays. Plant J 14: 643-652.

Dudek, D.J. 1989. Climate change impacts upon agriculture and resources: A case study of California. In: *The Potential Effects of Global Climate Change on the United States.* Appendix C, Agriculture, Vol. 1. EPA-230-05-89-053. J.B. Smith and D.A. Tirpak (eds.). US Environmental Protection Agency, Washington DC. pp. 5-1 to 5-38.

Ellstrand N. C. H. C. Prentice J. F. Hancock 1999 Gene flow and introgression from domesticated plants into their wild relatives. *Annual Review of Ecology and Systematics* 30: 539-563.

EU sponsored research on safety of Genetically Modified Organisms; DG Research 2001; http://europa.eu.int/comm/research/quality-of- life/gmo/index.html

Evans LT (1993) Crop Evolution, Adaptation and Yield. Cambridge University Press,

Cambridge, UK Farmelo G (2000) Why should scientists bother talking to the public? Science Next Wave. http://intl-nextwave.sciencemag.org/cgi/content/full/2000/01/06/5 (January 7, 2000)

FAO/WHO; January 2001; Evaluation of allergenici- ty of genetically modified food – report of a joint FAO/WHO Expert Consultation on Allergenicity of Foods derived from Biotechnology

FAO/WHO; June 2000; Safety aspects of geneti- cally modified foods of plant origin — Report of a Joint FAO/WHO Expert Consultation on Foods Derived from Biotechnology

Finch, C.V. and C.W. Sharp. 1976 Cover Crops in California Orchards and Vineyards. USDA Soil Conservation Service, Washington, D.C.

Fowler, C. And P. Mooney 1990. Shattering: food, politics and the loss of genetic diversity. University of Arizona Press, Tucson.

Francis, C.A. 1986. Multiple cropping systems. MacMillan, New York.

Fronwald S. S. Straus 1998 Genetically engineered oilseed rape (Agrevo/PGS): a critical assessment and background information. Greenpeace International, Amsterdam, Netherlands

Fry, G. 1995. Landscape ecology of insect movement in arable ecosystems. IN: Ecology and integrated farming systems. D.M. Glen et al. (eds). John Wiley and Sons, Bristol, UK.

Futuyama D. 1998 Evolutionary biology, 3rd ed. Sinauer, Sunderland, Massachusetts, USA

Gale MD, Devos KM (1998) Comparative genetics in the grasses. Proc Natl Acad Sci USA 95: 1971-1974.

Gianessi LP, Carpenter JE (1999) Agricultural biotechnology: insect control benefits. Report of the National Center for Food and Agricultural Policy, Washington, DC. http://www.bio.org/food&ag/bioins01.doc (July 1999)

Gladwin, C. anf K. Truman 1989. Food and farm: current debates and policies. University Press of America, Lantham, MD.

Gliessman, S.R. 1997. Agroecology: ecological processes in agriculture. Ann Arbor Press, Michigan.

Gliessman, S.R. 1998. Agroecology: ecological process in sustainable agriculture. Ann Arbor Press, Michigan.

Gliessman, S.R., E. Garcia and A. Amador. 1981. The ecological basis for the application of traditional agricultural technology in the management of tropical agro-ecosystems. Agro-Ecosystems 7:173-185.

Gressel J. 2002 Molecular biology of weed control. Taylor and Francis, London, UK

Gresshoff, P.M. 1996. Technology transfer of plant biotechnology. CRC Press, Boca Raton.

Grigg, D.B. 1974. The Agricultural Systems of the World: an evolutionary approach. Cambridge University Press, Cambridge.

Hall, A.E. and Allen, L.H. Jr. 1993. Designing cultivars for the climatic conditions of the next century. In: *International Crop Science I.* D.R. Buxton, R. Shibles, R.A. Forsberg, B.L. Blad, K.H. Asay, G.M. Paulsen and R.F. Wilson (eds.). Crop Science Society of America, Madison, Wisconsin. pp. 291-297.

Hammond-Kosack KE, Jones JDG (1997) Plant disease resistance genes. Annu Rev Plant Physiol Plant Mol Biol 48: 575-607.

Hansen, J., Fung, I., Lacis, A., Lebedeff, S., Rind, D., Ruedy, R., Russell, G. and Stone, P. 1988. Global climate changes as forecast by the GISS 3-D model. *J. Geophys. Res.* 98: 9341-9364.

Harwood, R.R. 1979. Small Farm Development — Understanding and Improving Farming Systems in the Humid Tropics. Westview Press, Boulder.

Hecht, S.B. 1984. Indigenous soil management in the Amazon basin: some implications for development In J.O. Browder (ed) Fragile Lands of Latin America pp. 166-181. Westview Press, boulder.

Hendrix, P.H., D.A.Jr. Crossley and D.C. Coleman. 1990. Soil biota as components of sustainable agroecosystems. In C.A. Edwards, Lal, Rattan, P. Madden, R. Miller, H. House, Gar (eds.) Sustainable Agricultural Systems. Soil and Water Conservation Society, IA.

Herrera-Estrella L (1999) Transgenic plants for tropical regions: some considerations about their development and their transfer to the small farmer. Proc Natl Acad Sci USA 96: 5978-5981.

Hindmarsh, R. 1991. The flawed "sustainable" promise of genetic engineering. The Ecologist 21: 196-205.

Ichida AM, Pei ZM, Baizabal-Aguirre VM, Turner KJ, Schroeder JI (1997) Expression of a Cs+-resistant guard cell K+ channel confers Cs+-resistant, light-induced stomatal opening in transgenic Arabidopsis. Plant Cell 9: 1843-1857.

Idso, S.B., Kimball, B.A., Anderson, M.G. and Mauney, J.R. 1987. Effects of atmospheric CO2 enrichment on plant growth: The interactive role of air temperature. *Agric. Ecosystems Environ.* 20: 1-10.

James C (1997) Progressing Public-Private Sector Partnerships in International Agricultural Research and Development. ISAAA Briefs No. 4. ISAAA, Ithaca, NY, pp 31

James C (1998) Global Review of Commercialized Transgenic Crops: 1998. ISAAA Briefs, No. 8. ISAAA, Ithaca, NY

James C. 2003 Preview: global status of commercialized transgenic crops: 2003. *ISAAA Briefs No. 30.* ISAAA, Ithaca, New York, USA

James, C. 1997. Global status of transgenic crops in 1997. ISAA Briefs, Ithaca, N.Y.

James, R.R. 1997. Utilizing a social ethic toward the environment in assessing genetically engineered insect-resistance in trees. Agriculture and Human Values 14: 237-249.

Jimenez-Osormio, J. and S. del Amo 1986. An intensive Mexican traditional agroecosystem: the chiampa. Proc. 6th International Scientific Conference IFOAM, Santa Cruz, California.

Jones, P., Allen, L.H. Jr. and Jones, J.W. 1985a. Responses of soybean canopy photosynthesis and transpiration to whole-day temperature changes in different CO2 environments. *Agron. J.* 77: 242-249.

Jones, P., Allen, L.H. Jr., Jones, J.W. and Valle, R.R. 1985b. Photosynthesis and transpiration responses of soybean canopies to short- and long-term CO2 treatments. *Agron. J.* 77: 119-126.

Jones, P., Allen, L.H. Jr., Jones, J.W. Boote, K.J. and Campbell, W.J. 1984. Soybean canopy growth, photosynthesis, and transpiration responses to whole-season carbon dioxide enrichment. *Agron. J.* 76: 633-637.

Jones, P., Jones, J.W. and Allen, L.H. Jr. 1985c. Carbon dioxide effects on photosynthesis and transpiration during vegetative growth in soybeans. *Soil Crop Sci. Soc. Fla. Proc.* 44: 129-134.

Karp A, Edwards KJ, Bruford M, Funk S, Vosman B, Morgante M, Seberg O, Kremer A, Boursot P, Arctander P, Tautz D, Hewitt GM (1997) Molecular technologies for biodiversity evaluation: opportunities and challenges. Nat Biotech 15: 625-628.

Keeling, C.D., Whorf, T.P, Wahlen, M. and van der Plicht, J. 1995. Interannual extremes in the rate of rise of atmospheric carbon dioxide since 1980. *Nature* 375: 660-670.

Khan MS, Maliga P (1999) Fluorescent antibiotic resistance marker for tracking plastid transformation in higher plants. Nat Biotech 17: 910-915.

Kimball, B.A. 1983. Carbon dioxide and agricultural yield: An assemblage and analysis of 430 prior observations. *Agron. J.* 75: 779-788.

Kimball, B.A., Mauney, J.R., Nakayama, F.S. and Idso, S.B. 1993. Effects of increasing atmospheric CO2 on vegetation. *Vegetation* 104/105: 65-75. Also: pp. 65-75. In: CO2 *and Biosphere.* (Advances in Vegetation Science 14). J. Rozema, H. Lambers, S.C. van de Geijn and M.L. Cambridge (eds.). Kluwer Academic Publishers, Dordrecht.

Kleinman, D.L. and J. Kloppenburg 1988. Biotechnology and university-industry relations: policy issues in research and the ownership of intellectual property at a land grant university. Policy Studies Journal 17: 83-96.

Kloppenburg, J.R. 1988. First the seed: the political economy of plant technology, 1492-2000. Cambridge University Press, Cambridge.

Koo M, Bendahmane M, Lettieri GA, Paoletti AD, Lane TE, Fitchen JH, Buchmeier MJ, Beachy RN (1999) Protective immunity against murine hepatitis virus (MHV) induced by intranasal or subcutaneous administration of hybrids of tobacco mosaic virus that carries an MHV epitope. Proc Natl Acad Sci USA 96: 7774-7779.

Kramer, P.J. 1981. Carbon dioxide concentration, photosynthesis, and dry matter production. *Bioscience* 31: 29-33.

Krimsky, S. and R.P. Wrubel 1996. Agricultural biotechnology and the environment: science, policy and social issues. University of Illinois Press, Urbana.

Kumagai M, Donson J, Della-Cioppa G, Harvey D, Hanley K, Grill L (1995) Cytoplasmic inhibition of carotenoid biosynthesis with virus-derived RNA. Proc Natl Acad Sci USA 92: 1679-1683.

Kunkel T, Niu QW, Chan YS, Chua NH (1999) Inducible isopentenyl transferase as a high-efficiency marker for plant transformation. Nat Biotech 17: 916-919.

Lentz, D.I. 1986. Ethnobotany of the Jicaque of Honduras. Econ. Bot. 40: 210-219

Levidow, L. and S. Carr 1997. How biotechnology regulation sets a risk / ethics boundary. Agriculture and Human Values 14: 29-43.

Liebman, J. 1997. Rising toxic tide: pesticide use in California, 1991-1995. Report of Californians for Pesticide Reform and Pesticide Action Network. San Francisco.

Liebman, M. and T. Ohno. 1998. Crop rotation and legume residue effects on weed emergence and growth: implications for weed management. In J.L. Hotfield and B.A. Stwerrt (eds.). Integrated weed and soil management. Ann Arbor Press, MI

Lin XY, Kaul SS, Rounsley S, Shea TP, Benito MI, Town CD, Fujii CY, Mason T, Bowman CL, Barnstead M, Feldblyum TV, Buell CR, Ketchum KA, Lee J, Ronning CM, Koo HL, Moffat KS, Cronin LA, Shen M, Pai G, VanAken S, Umayam L, Tallon LJ, Gill JE, Adams MD, Carrera AJ, Creasy TH, Goodman HM, Somerville CR, Copenhaver GP, Preuss D, Nierman WC, White O, Eisen JA, Salzberg SL, Fraser CM, Venter JC (1999) Sequence and analysis of chromosome 2 of the plant *Arabidopsis thaliana*. Nature 402: 761-768.

Lipton, M. 1989. New seeds and poor people. The John Hopkins University Press, Baltimore.

Lloyd JR, Springer F, Buleon A, Muller-Rober B, Willmitzer L, Kossmann J (1999) The influence of alterations in ADP-glucose pyrophosphorylase activities on starch structure and composition in potato tubers. Planta 209: 230-238.

Long, S.P. 1991. Modification of the response of photosynthetic productivity to rising temperature by atmospheric CO2 concentrations: Has its importance been underestimated? *Plant Cell Environ.* 14: 729-739.

Losey JE, Rayor LS, Carter ME (1999) Transgenic pollen harms monarch larvae. Nature 399: 214.

Lutman P. [ed.] 1999 Gene flow and agriculture. Relevance for transgenic crops. *BCPC Symposium No. 72.* British Crop Protection Council, Farnham, UK

Ma JKC, Hikmat BY, Wycoff K, Vine ND, Chargelegue D, Yu L, Hein MB, Lehner T (1999) Characterization of a recombinant plant monoclonal secretory antibody and preventive immunotherapy in humans. Nat Med 4: 601-606

MacDonald, D.F. 1991. Agricultural biotechnology at the crossroads. NABC Report 3. Union Press of Binghamton.

MacDonald, J.F. 1994. Agricultural biotechnology and the public good. NABC Report 6. Ithaca, NY.

Mae-Wan Ho; May 2002; Frequently Asked Questions on GMOs; http://www.i-sis.org.uk/FAQ.php

Maliga P, Nixon PJ (1998) Judging the homoplastomic state of plastid transformants. Trends Plant Sci 3: 376-377.

Manabe, S. and Wetherald, R.T. 1987. Large scale changes of soil wetness induced by an increase in atmospheric carbon dioxide. *J. Atmos. Sci.* 44: 1211-1235.

Mander, J. And E. Goldsmith 1996. The case against the global economy. Sierra Club Books, San Francisco.

Marten, G.C. 1986. Traditional agriculture in South-east Asia: a human ecology perspective. Westview Press, Boulder.

Mayer K, Schuller C, Wambutt R, Murphy G, Volckaert G, Pohl T, Dusterhoft A, Stiekema W, Entian KD, Terryn N, Harris B, Ansorge W, Brandt P, Grivell L, Rieger M, Weichselgartner M, de Simone V, Obermaier B, Mache R, Muller M, Kreis M, Delseny M, Puigdomenech P, Watson M, McCombie WK (1999) Sequence and analysis of chromosome 4 of the plant *Arabidopsis thaliana*. Nature 402: 769-777.

Mc Guinnes, H. 1993. Living soils: sustainable alternatives to chemical fertilizers for developing countries. Unpublished manuscript, Consumers Policy Institute, New York.

Mc Isaac, G. and W.R. Edwards 1994. Sustainable agriculture in the American midwest. University of Illinois Press, Urbana.

McHughen, Alan. *Pandora's Picnic Basket: The Potential Hazards of Genetically Modified Foods.* New York: Oxford University Press, 2000. Nuffield Council of Bioethics. *Genetically Modified Crops: The Ethical and Social Issues.* London: Nuffield Council, 1999.

Michael A., John A. Hillman, and David J. Robinson. "Genetic Modification in Context and Perspective." *Fearing Food: Risk, Health & Environment.* Edited by Julian Morris and Roger Bate. Boston: Butterworth Heinemann, 1999, 58-77.

Miflin BJ (2000) Crop improvement in the 21st century. J Exp Bot 51: 1-8.

Molnar, J.J. and H. Kinnucan 1989. Biotechnology and the new agricultural revolution. Westview Press, Boulder, CO.

Munkvold GP, Hellmich RL, Rice LG (1999) Comparison of fumonisin concentrations in kernels of transgenic Bt maize hybrids and nontransgenic hybrids. Plant Dis 83: 130-138.

Nair, P.K.R. 1982. Soil Productivity Aspects of Agroforestry. ICRAF, Nairobi.

Nair, P.K.R. 1984. Soil productivity aspects of agroforestry. ACRAF, Nairobi.

National Research Council 1996. Ecologically based pest management. National Academy of Sciences. Washington D.C.

Noel L, Moores TL, vanderBiezen EA, Parniske M, Daniels MJ, Parker JE, Jones JDG (1999) Pronounced intraspecific haplotype divergence at the RPP5 complex disease resistance locus of *Arabidopsis*. Plant Cell 11: 2099-2111.

Norman, M.J. T. 1979. Annual Cropping Systems in the Tropics. University Presses of Florida, Gainesville.

Noteborn, H P J M, Bienenmann-Ploum, M E, Van den Berg, J H J, Alink, G M, Zolla, L, Reynaerts, A, Pensa, M, Kuiper, H A; 1995; Safety assessment of the Bacillus thuringiensis insecticidal crystal pro- tein Cry1A(b) expressed in transgenic tomatoes; in Genetically Modified Foods – safety issues; ACS Symposium Series 605, American Chemical Society, Washington DC pp 134-147

Office of Technology Assesment 1992. A new technological era for American agriculture. U.S. Government Printing Office. Washington. D.C.

Parniske M, Jones JDG (1999) Recombination between diverged clusters of the tomato Cf-9 plant disease resistance gene family. Proc Natl Acad Sci USA 96: 5850-5855.

Parry, M.L., Carter, T.R. and Konijn, N.T. (eds.). 1988. *The Impact of Climatic Variations on Agriculture.* Vols. 1 and 2. Kluwer Academic Publishers, Dordrecht.

Pearson, C.J. and R.L. Ison. 1987. Agronomy of grassland systems. Cambridge University Press, Cambridge.

Peart, R.M., Jones, J.W., Curry, R.B., Boote, K.J. and Allen, L.H. Jr. 1989. Impact of climate change on crop yield in the Southeastern U.S.A.: A simulation study. In: *The Potential Effects of Global Climate Change on the United States.* Appendix C, Agriculture, Vol. 1, EPA-230-05-89-053. J.B. Smith and D.A. Tirpak (eds.). US Environmental Protection Agency, Washington DC. pp. 2-1 to 2-54.

Pickering, N.B., Jones, J.W. and Boote, K.J. 1995. Adapting SOYGRO V5.42 for prediction under climate change conditions. In: *Climate Change and Agriculture: Analysis of Potential International Impacts.* C. Rosenzweig, L.A. Harper, S.E. Hollinger, J.W. Jones and L.H. Allen, Jr. (eds.). ASA Special Pub. No. 59, American Society of Agronomy, Madison, Wisconsin. pp. 77-98.

Pimentel, D. and H. Lehman 1993. The pesticide question. Chapman and Hall, N.Y.

Pimentel, D. and M. Pimentel 1979. Food, Energy and Society. Edward Arnold, London.

Pimentel, D. et al. 1992. Environmental and economic costs of pesticide use. Bioscience 42: 750-760.

Pohl-Orf M, Brand U, Driessen S, Hesse PR, Lehnen M, Morak C, Mucher T, Saeglitz C, von Soosten C, Bartsch D (1999) Overwintering of genetically modified sugar beet, *Beta vulgaris* L. subsp *vulgaris*, as a source for dispersal of transgenic pollen. Euphytica 108: 181-186.

Poorter, H. 1993. Interspecific variation in the response of plants to an elevated ambient CO2 concentration. *Vegetation* 104/105: 77/97. Also: pp. 77-97. In: CO2 *and Biosphere.* (Advances in Vegetation Science 14). J: Rozema, H. Lambers, S.C. van de Geijn and M.L. Cambridge (eds.). Kluwer Academic Publishers, Dordrecht.

Pretty, J. 1995. Regenerating agriculture. World Resources Institute. Washington, DC.

Pretty, J. 1997. The sustainable intensification of agriculture. Natural Resources Forum 21: 247- 256.

Pretty, J.N. 1994. Regenerating Agriculture. Earthscan Publications Ltd., London.

Pusztai, A; October 1998; Report of Project Coordinator on data produced at the Rowett Research Institute; http://www.rowett.ac.uk/gmo/

RCEP (Royal Commission on Environmental Pollution). 1989 The release of genetically engineered organisms into the environment. 13th Report, Royal Commission on Environmental Pollution. HMSO, London, UK

Reijntjes, C.B., Haverkort and A. Waters-Bayer. 1992. Farming for the future. MacMillan Press Ltd., London.

Reinjtes, C., B. Haverkort and Ann Waters-Bayer 1992. Farming for the future. MacMillan, London.

Richards, P. 1985. Indigenous Agricultural Revolution. Westview Press, Boulder.

Rissler, J. and M. Mellon 1996. The ecological risks of engineered crops. MIT Press, Cambridge.

Ritchie, J.T., Baer, B.D. and Chou, T.Y. 1989. Effect of global climate change on agriculture: Great Lakes Region. In: *The Potential Effects of Global Climate Change on the United States.* Appendix C, Agriculture, Vol. 1. EPA-230-05-89-053. J.B. Smith and D.A. Tirpak (eds.). US Environmental Protection Agency, Washington DC. pp. 1-1 to 1-42.

Robinson, R.A. 1996. Return to resistance: breeding crops to reduce pesticide resistance. AgAccess, Davis, CA.

Rosenzweig, C. 1989. Potential effects of climate change on agricultural production in the Great Plains: A simulation study. In: *The Potential Effects of Global Climate Change on the United States.* Appendix C, Agriculture, Vol. 1. EPA-230-05-89-053. J.B. Smith and D.A. Tirpak (eds.). US Environmental Protection Agency, Washington DC. pp. 3-1 to 3-43.

Rosenzweig, C. and Iglesias, A. (eds.). 1994. *Implications of Climate Change for International Agriculture: Crop Modelling Study.* EPA 230-B-94-003. US Environmental Protection Agency, Washington DC.

Rosenzweig, C. and Parry, M.L. 1993. Potential impact of climate change on world food supply: A summary of a recent international study. In: *Agricultural Dimensions of Global Climate Change.* H.M. Kaiser and T.E. Drennen (eds.). St. Lucie Press, Delray Beach, Florida, pp. 87-116.

Rosenzweig, C. and Parry, M.L. 1994. Potential impact of climate change on world food supply. *Nature* 367: 133-138.

Rosenzweig, C., Harper, L.A., Hollinger, S.E., Jones, J.W. and Allen, L.H. Jr. (eds.). 1995. *Climate Change and Agriculture: Analysis of Potential International Impacts.* ASA Special Pub. No. 59, American Society of Agronomy, Madison, Wisconsin.

Rosset, P.M. and M.A. Altieri 1997. Agroecology versus input substitution: a fundamental contradiction in sustainable agriculture. Society and Natural Resources 10: 283-295.

Rowland-Bamford, A.J., Allen, L.H. Jr., Baker, J.T. and Bowes, G. 1991. Acclimation of rice to changing atmospheric carbon dioxide concentration. *Plant Cell Environ.* 14: 577-583.

Rozema, J., Lambers, H., van de Geijn, S.C. and Cambridge, M.L. (eds.). 1993. *CO2 and Biosphere.* (Advances in Vegetation Science 14). Kluwer Academic Publishers, Dordrecht.

Sanders, W.T. 1957. Tierra y agua: a study of the ecological factors in the development of Meso- American civilizations. PhD Dissertation, Harvard University.

Satake, T. and Yoshida, S. 1978. High temperature induced sterility in indica rices at flowering. *Japan. J. Crop Sci.* 47: 6-17.

Schuler TH, Poppy GM, Kerry BR, Denholm I (1998) Insect-resistant transgenic plants. Trends Biotech 16: 168-175.

Smith, J.B. and Tirpak, D.A. (eds.). 1989. *The Potential Effects of Global Climate Change on the United States.* EPA-230-05-89-050. US Environmental Protection Agency, Washington DC.

Somerville CR, Ogren WL (1979) A phosphoglycolate phosphatase-deficient mutant of *Arabidopsis*. Nature 280: 833-836.

Stace C. A. 1975 Hybridization and the flora of the British Isles. Academic Press, London, UK

Stewart C. N. M. D. Halfhill S. I. Warwick 2003 Transgene introgression from genetically modified crops to their wild relatives. *Nature Reviews Genetics* 4: 806-817.

Sumner, D.R. 1982. Crop rotation and plant productivity. In M. Recheigl, (ed). CRC Handbook of Agricultural Productivity, Vol. I CRC Press, Florida.

Swift, M.S., J. Vandermeer, P.S. Ramakrishnan, J.M. Anderson, C.K. Ong and B.A. Hawkins 1996. Biodiversity and agroecosystem function. H.A. Mooney et al. (eds.). Functional roles of biodiversity: a global perspective. J. Wiley and Sons, N.Y., pp. 261-298.

Tanksley SD, McCouch SR (1997) Seed banks and molecular maps: unlocking genetic potential from the wild. Science 277: 1063-1066.

Thies, C. and T. Tscharntke. 1999. Landscape structure, and biological control in agroecosystems. Science 285: 893-895.

Thompson, P.B. 1995. The spirit of the soil: agriculture and environmental ethics. Routledge, London.

Thorne, J.M. and Koller, R.M. 1974. Influence of assimilate demand on photosynthesis, diffusive resistance, translocation and carbohydrate levels of soybean leaves. *Plant Physiol.* 54: 201-207.

Thrupp, L.A. 1998. Cultivating diversity: agrobiodiversity and food security. World Resources Institute, Washington, DC.

Thurston, H.D. 1991. Sustainable practices for plant disease management in traditional farming systems. Westview Press, Boulder.

Tilman, D, D. Wedin and J. Knops 1996. Productivity and sustainability influenced by biodiversity in grassland ecosystems. Nature 379: 718-720.

Toledo, V.M. 1995. Peasantry, agroindustriality and sustainability. Interamerican Council for Sustainable Agriculture. Working Paper 3. Michoacan, Mexico.

Toledo, V.M., J. Carabias, C. Mapes, and C. Toledo. 1985. Ecologia y Autosuficiencia Alimentaria. Siglo Vientiumo Editors, Mexico City.

UNDP 1995. Benefits of Diversity. UNDP, New York. Vandermeer, J. and I. Perfecto 1995. Breakfast of biodiversity: the truth about rainforest destruction. Food First Books, Oakland.

Valle, R., Mishoe, J.W, Campbell, W.J., Jones, J.W. and Allen, L.H. Jr. 1985. Photosynthetic responses of 'Bragg' soybean leaves adapted to different CO2 environments. *Crop Sci.* 25: 333-339.

Van Dyke H. P. Boudry H. McCombie P. Vernet 1997 Flowering time in wild beet (*Beta vulgaris* ssp. *maritima*) along a latitudinal cline. *Acta Oecologia* 18: 47-60

Vandermeer, J. 1989. The ecology of intercropping. Cambridge University Press, Cambridge.

Vandermeer, J. 1995. The ecological basis of alternative agriculture. Annual Review of Ecological Systems 26: 201-224.

Veramendi J, Roessner U, Renz A, Willmitzer L, Trethewey RN (1999) Antisense repression of hexoki nase 1 leads to an overaccumulation of starch in leaves of transgenic potato plants but not to significant changes in tuber carbohydrate metabolism. Plant Physiol 121: 123-133.

Vergara, B.S. 1980. Rice plant growth and development. In: *Rice Production and Utilization.* B.S. Luh (ed.). Avi Publishing, Westport, CT. pp. 75-86.

Vuylsteke M, Mank R, Antonise R, Bastiaans E, Senior ML, Stuber CW, Melchinger AE, Lubberstedt T, Xia XC, Stam P, Zabeau M, Kuiper M (1999) Two high-density AFLP(R) linkage maps of *Zea mays* L.: analysis of distribution of AFLP markers. Theor Appl Genet 99: 921-935.

Wambugu F (1999) Why Africa needs agricultural biotech. Nature 400: 15-16.

Warwick S. I. M. J. Sumard A. Legere H. J. Beckie B. Zhu P. Mason G. Seguin-Swartz C. N. Stewart 2003 Hybridization between transgenic *Brassica napus* L. and its wild relatives: *Brassica rapa* L., *Raphanus raphanistrum* L., *Sinapis arvensis* L., and *Erucastrum gallicum* (Willd.) O.E. Schulz. *Theoretical and Applied Genetics* 107: 528-539.

Webber, D.J. (ed) 1990. Biotechnology: assessing social impacts and policy implications. Greenwood Press, Westport, CT.

Wilken, G.C. 1987. Good Farmers: traditional agricultural resource management in Mexico and Guatemala. University of California Press, Berkeley.

Wilson, C.A. and Mitchell, J.F.B. 1987. A doubled CO2 climate sensitivity experiment with a global model including a simple ocean. *J. Geophys. Res.* 92: 13315-13343.

Windham GL, Williams WP, Davis FM (1999) Effects of the southwestern corn borer on *Aspergillus flavus* kernel infection and aflatoxin accumulation in maize hybrids. Plant Dis 83: 535-540.

Wittwer, S.H. 1995. *Food, Climate, and Carbon Dioxide: The Global Environment and World Food Production.* CRC Press/Lewis Publishers, Boca Raton, Florida.

Woodwell, G.M. and Mackenzie, F.T. (eds.). 1995. *Biotic Feedbacks in the Global Climatic System: Will the Warming Increase the Warming?* Oxford University Press, New York.

Wullschleger, S.D., Post, W.M. and King, A.W. 1994. On the potential for a CO2 fertilization effect in forests: Estimates of the Biotic Growth Factor based on 58 controlled-exposure studies. In: *Biotic Feedbacks in the Global Climatic System: Will the Warming Increase the Warming?* G.M. Woodwell and F.T. Mackenzie (eds.). Oxford University Press, New York. pp. 85-107.

Xiao JH, Li JM, Grandillo S, Ahn SN, Yuan LP, Tanksley SD, McCouch SR (1998) Identification of trait-improving quantitative trait loci alleles from a wild rice relative, *Oryza rufipogon*. Genetics 150: 899-909.

Ye X, Al-Babili S, Klöti A, Zhang J, Lucca P, Beyer P, Potrykus I (2000) Engineering the provitamin a (-carotene) biosynthetic pathway into (carotenoid-free) rice endosperm. Science 287: 303-305

Yoshida, S. 1977. Rice. In: *Ecophysiology of Tropical Crops.* P.T. Alvim and T.T. Kozlowski (eds.). Academic Press, London. pp. 57-87.

Yoshida, S. 1981. *Fundamentals of Rice Crop Science.* International Rice Research Institute, Los Baños, Philippines.

Index

H

I

J

K

L

M

■■■